Praise for Backbone

"OUTSTANDING! A great capture of the Marine NCO. This should be required reading for all Devil Dogs. A big 'atta boy' goes to the author. Any businessman, or anyone in a leadership position, will benefit greatly by reading this book."

—GySgt R. Lee Ermey

"This remarkable book cracks the bone and examines the 'marrow' of leadership. One of the best books ever written about leadership, with powerful, riveting examples and direct application to the business world, or any other field of endeavor. Required reading for every Marine, for everyone who honors and studies our military, and for everyone who studies and strives in the field of leadership."

—LtCol Dave Grossman, USA (ret.), author of
On Killing and *On Combat*

"Dr. Julia D. Dye has written an absolutely fascinating book on U.S. Marine Corps noncommissioned officers (NCOs)—the backbone of the Corps. Taking fourteen leadership traits that are endemic to what constitutes the essence of a Marine NCO, Dr. Dye deftly interweaves personal stories from the past and present and demonstrates why the Marine Corps remains the elite military service that it is today. I predict that Dr. Dye's book will be required reading for anyone who aspires to improve themselves or, more importantly, desires to wear the earned and respected chevrons of a Marine NCO."

—Dr. Charles P. Neimeyer, Director and Chief, USMC History

"*Backbone* is essential reading for everyone in a position of leadership—and all who aspire to lead—no matter their walk of life. Julia Dye nails it—and her exemplars are the finest leaders on the planet – people who know the meaning of 'Semper Fidelis.'"

—LtCol Oliver North, USMC (Ret.), Host of "War Stories"
on FOX News Channel

"*Backbone* defines the spirit and ethos of leadership exemplified by the few and the proud, the United States Marine. Julia Dye focuses on the critical role of the noncommissioned officer as the nucleus for successful mission accomplishment, she gives the reader a fascinating insight to the culture, ethics, and tradition that has molded Marine NCOs as leaders on and off the battlefield for generations. Backbone gives us an understanding of the unique metrics that shape Marine NCOs, giving the reader a compelling overview on what defines, influences, and creates successful leaders. *Backbone* is a must read for any organization or individual that views leadership as being centric to their success"

—Captain Jim Palmersheim, Managing Director, Veterans and
Military Programs, American Airlines

"Few people have the privilege of serving as a noncommissioned officer in the U.S. Marine Corps, but anyone who is interested in leading the kind of purpose-driven and values-centered life that marks the best Marine NCOs can learn from this book. And anyone in the business of leading people—whether it's in business, the classroom, or the playing field—should study the men and women Dye writes about with such insight. These NCOs don't just teach young Marines; they have a great deal to teach anyone smart enough to read, listen, and learn."

—Ed Ruggero, author of *The First Men In: US Paratroopers
and the Fight to Save D-Day*

"A primer for life and a classic about values. While certain to be studied in military circles, it actually deserves to be read by all college students."

—Bing West, best-selling author of *The Village,
The Strongest Tribe,* and *The Wrong War*

Backbone

History, Traditions, and
Leadership Lessons of
Marine Corps NCOs

Julia Dye, Ph.D.

WARRIORS PUBLISHING GROUP

NORTH HILLS, CALIFORNIA

BACKBONE: History, Traditions, and Leadership Lessons of Marine Corps NCOs

A Warriors Publishing Group book/published by arrangement with the author.

PRINTING HISTORY
Warriors Publishing Group edition/October 2012
Previously published by Osprey Publishing

Every attempt has been made by the Publisher to secure the appropriate permissions for material reproduced in this book. If there has been any oversight we will be happy to rectify the situation in future editions and written submission should be made to the Publisher.

The views expressed by the parties named herein do not necessarily reflect the official views or policies of the United States Marine Corps.

ISBN 978-0-9897983-8-9

The name "Warriors Publishing Group" and the logo
are trademarks belonging to Warriors Publishing Group

PRINTED IN THE UNITED STATES OF AMERICA

10 9 8 7 6 5 4 3 2 1

Front cover: Regimental Combat Team 8 Marines stand on a line prepared to fire at targets during a combat marksmanship program held on Range Panel Stages aboard Marine Corps Air Station Yuma, Arizona, in October 2010. (Photo by Lance Corporal Clayton L. VonDerAhe.)

For my father,
who always believed

Acknowledgments

S ome special thanks are in order here. No one could possibly write a book of this nature alone. My extreme gratitude goes out first to the men and women of the United States Marine Corps who continue to serve this nation with grace and courage. Special thanks go out to those Marines who allowed me to speak with them and discover their thoughts and experiences with leadership in real-world situations.

To Major Mark Shuster: as my research assistant and liaison with the Marine Corps, your tireless efforts to build bridges between Marines and civilians are inspirational.

To the United States Marine Corps for all the support in allowing access to the NCOs whose voices are heard throughout this book. In particular, special gratitude for the assistance of Lieutenant Colonel Joseph Clearfield, Sergeant Major Ramona Cook, Sergeant Major John Cook, Sergeant Major Shelley Sergeant, Major Carl Redding, Jr., Captain Mike Alvarez, Gunnery Sergeant Chanin Nuntavong, and Gunnery Sergeant Sheryl Wilhoit. And from the U.S. Navy: LT Paul Macapagal, Lieutenant Doug Freudenberger, David P. German, and Michael McLellan, and finally Maxine MacGregor and Beth L. Crumley.

To my husband, Captain Dale A. Dye, USMC (Retired): You wield a keyboard with as much strength and skill as you wield a weapon. Thanks for your keen eye, your patience, and your support through it all. And for dog walks and tolerating one more pizza night!

Foreword

I am the backbone of the United States Marine Corps, I am a Marine Non-Commissioned Officer. I serve as part of the vital link between my commander (and all officers) and enlisted Marines. I will never forget who I am or what I represent. I will challenge myself to the limit and be ever attentive to duty. I am now, more than ever, committed to excellence in all that I do, so that I can set the proper example for other Marines. I will demand of myself all the energy, knowledge and skills I possess, so that I can instill confidence in those I teach. I will constantly strive to perfect my own skills and to become a good leader. Above all I will be truthful in all I say or do. My integrity shall be impeccable as my appearance. I will be honest with myself, with those under my charge and with my superiors. I pledge to do my best to incorporate all the leadership traits into my character. For such is the heritage I have received from that long, illustrious line of professionals who have worn the bloodstripe so proudly before me. I must give the very best I have for my Marines, my Corps and my Country for though today I instruct and supervise in peace, tomorrow, I may lead in war.[1]

Certain institutions carry a certain type of gravitas. Those who can claim to be a part of that institution carry that bearing with them. It can be seen in the way they walk, in the confidence of their speech, in the respect they give others. Of American institutions, the Marine Corps ranks among the very best: Harvard. The Metropolitan Opera. The Smithsonian Institution. NASA. The Mayo Clinic. Only the best get in. Each represents the pinnacle, the top of their industry—the very best of our meritocracy.

The Marine Corps is the only branch of the American Armed Forces that recruits people specifically to fight. Other branches may promise education benefits, travel, personal development, and fun. The Marine Corps promises only the honor of being a United States Marine.

A look at the recruiting websites exemplifies these differences. The Marine Corps site states:

"Earned. Never Given. We don't ask for anything more than everything you've got .... You will not be given anything other than the opportunity to prove that you have the courage to stand on an impenetrable line of warriors stretching 234 years. Our title is earned, never given. And what's earned is yours forever."

That's why traditionally there are no Ex-Marines. Once a Marine, always a Marine.

Compare that to the Army's site: "With the skills and training you gain in the Army, there's no limit to how far you can go."

Or the Air Force: "Healthcare, travel, leadership skills, housing, and of course a pay check. The benefits of an Air Force career are many."

The other service recruiters focus on what they can do for the men and women who join. The Marine Corps focuses on what you can do for the Corps: something larger and more important than yourself.

And Marine Corps training is more challenging—physically and mentally—than the basic training programs of any of the other military services. Not only are the physical requirements much higher, but recruits are required to learn and memorize a startling amount of information, including the history and traditions of the Marine Corps. At about twelve and a half weeks, it's also the longest.

U.S. Marines with India Company, 3rd Battalion, 5th Marines provide security in the Sangin valley, Helmand province, Afghanistan, Dec. 6, 2010. The battalion was part of Regimental Combat Team 2 which conducted counterinsurgency operations in partnership with the International Security Assistance Force. (U.S. Marine Corps photo by Cpl. David Hernandez/ Released.)

The Marine Corps is the smallest of the armed forces in the U.S. Department of Defense.[2] As of September 30, 2010, there were just under 202,500 active duty Marines, and that's at wartime strength.[3] When the country is not at war, levels generally are well under 200,000.

This smaller size allows the Marine Corps to be more familial than the other branches. Over the course of a career, a Marine personally knows a high percentage of fellow Marines. If a Marine is struggling, others are there to support him and to help ensure his success. Although the perception of the Marines is one of rigidity and uniformity, in reality the Marine Corps allows for diverse personalities and unique career paths.

Part of this unique nature is due to the Corps' amalgamation of fighting styles. Combining the best characteristics of soldiers, sailors, and airmen, the Marine is a sea soldier—an odd conglomeration that talks like one, dresses like another, and fights like them all. To be different, and to remain different, is important to Marines. This difference is expressed through strict obedience to orders, military appearance, disciplined behavior, and an unyielding conviction that they exist to fight. And since every Marine, enlisted or officer, goes through the same training experience, the Corps maintains a sense of cohesiveness like no other American service.

The cohesiveness is evident in a number of ways, not the least of which is through a Marine's uniform. In the Navy, Sailors wear rating badges that identify their jobs. A Soldier wears branch of service insignia on his collar, with metal shoulder pins and cloth sleeve patches to identify his unit. Marines, on the other hand, wear only the Eagle, Globe and Anchor, their ribbons, and their marksmanship badges. Just by looking at the uniform, you cannot tell what they do each day, nor the unit to which they belong. A Marine may drive AmTracs, program computers, or fly helicopters. The tasks are not important. What is important is that the Marine is a Marine.

This book is about the Corps—and the Noncommissioned Officers who serve: they are the core of the Corps.

Although commissioned officers are in charge of setting policy, NCOs apply that policy and make crucial judgments on the ground. The amount of trust given to NCOs is the key to success of this division of labor. Decisions and actions take place every day with no officer present. Decentralized, implicit trust lends a huge advantage to the Corps and makes the NCO a combat multiplier: a force that significantly increases the Corps' combat potential, thereby enhancing the probability of successful mission accomplishment.

Sergeants train individuals, teams, and crews. NCOs focus on all the single and small unit requirements that support the collective tasks of platoons and companies. They ensure that their fellow Marines are physically fit to arrive and succeed at the leading edge of battle.

NCOs also advise and mentor officers. From the platoon level, an officer and a seasoned NCO work together to accomplish missions. This unique combination of commissioned and noncommissioned officers is a powerful system of leadership—and it works.

Leaders influence people by providing purpose, direction, and motivation while operating to accomplish the mission and improve the organization. The Marine Corps believes this is possible only through the leader's adoption of specific character attributes or leadership traits. Dr. Steven M. Silver, a psychologist and former Marine who served in Vietnam, explains how the Marines Corps' approach to leadership sets it apart from other branches of service:

> "*Because of its reliance on small unit leadership and because of its expertise in unconventional warfare, no branch of the U.S. Armed Forces has placed a greater emphasis on manifestations of ethics—loyalty, integrity, courage, and honor—than the Marine Corps.*"[4]

These leadership traits—defined above as "ethics"—are reinforced to Marines at every level of their training. And good NCOs teach these traits every day through the example of their own leadership.

Marine Corps NCOs aren't overly-regimented parrots, mimicking orders down the line. They are characters and individuals who bring their own color to the job. Some are quiet, efficiently doing their job every day. Others are remembered and heralded throughout history.

This reverence for the NCO is a special trait of the Marine Corps. The Army has its heroes: Patton, MacArthur, Pershing. So does the Navy: Farragut, Nimitz, and Halsey. These men all have one thing in common—they were all officers. Ask a Marine about his or her Marine heroes, however, and you'll likely get a list like this: Lou , John Basilone, Carlos Hathcock, Dan Daly—all NCOs.

An NCO is intimately connected with the two precepts of military leadership: accomplish the mission and insure the welfare of your Marines. A Marine is a valuable commodity to be protected. To send untrained, undisciplined Marines into combat is to put their lives in danger.

The Marine Corps needs NCOs. Small unit leaders are trained, honored, and trusted. The corporate world needs small unit leaders, too. But we don't revere the assistant managers at Walgreens, for example. Those people are on the front lines, they're dealing with problems every day, they've got to motivate their people, and they have to deal with customers. But they need a certain respect, too.

Jerry Anderson, a Vietnam-era NCO, explains how today's businesses fail their leaders:

> *"Upper leaders won't let small unit leaders make decisions because they feel it will reflect poorly on them. They haven't either trained them properly, or given them the direction they need to make decisions that are in line with what the company wants."*[5]

Leaders are human, so they are going to make mistakes. That gives an opportunity for instruction so mistakes won't be repeated. And this continual training is vital. Sergeant Robert Bayer is another Vietnam veteran who transferred what he learned as a Marine NCO to his new career as an editor for the Los Angeles Times. He found that <u>it is important to prepare leaders</u> both in the Corps and in business:

IMPORTANT OF LDRSHP EDUCATION + TRNG

> *"So you have a guy who was a meat packer for years, knows how the job is done, the policies and procedures, and then they need a new meat packer supervisor, and he gets the job. He's now a 'corporal.' But even though he knows how they want the meat packed, he doesn't know about being a corporal. He doesn't have any training at all. He doesn't know how to deal with his new leadership challenges."*[6]

How can we provide small unit leaders with the appropriate training and direction to build that trust and allow them to make decisions? We can look to examples of excellent small unit leaders and see how they were taught, and how they are teaching others, through instruction and leadership by example.

No one does it better than the Marine Corps.

Marine Corps NCOs know that survival is crucial both on the battlefield and in the business arena. You can neither win nor succeed if you don't survive. Business leaders know this as well. Victory on the battlefield equates to success in the business battle—it's about the mission and keeping that mission central to each tactical decision.

This book will explore common concerns between NCOs and leaders in the civilian sectors. Things like leadership, training, teamwork, task organization, morale, technical proficiency and discipline are applicable regardless of the nature of the battle space.

While the civilian world focuses on management, the lessons of the NCO are about leadership. They are not the same thing. You can manage resources—but you must lead people.

Just as global military battles become increasingly non-traditional and asymmetrical, business is becoming more like guerilla warfare than ever before. No longer can a chief executive officer make all the decisions. Leaders must fight on many fast-changing fronts and it is impossible for them to be close to all of them. Today's complex and knowledge-intensive world requires the kind of bottom-up leadership that NCOs undertake every day. By encouraging front line troops to innovate and lead by trusting them to understand the mission and those people in their charge, small units can work independently in far-flung theatres of operation, maintaining their commander's intent in each of their decisions.

The American tradition of the NCO began early in our country's birth. In the early days of the American Revolution, little standardization of NCO duties or responsibilities existed. Like the American military itself, the NCO Corps blended traditions of the French, British, and Prussian armies into a uniquely American institution. As the years progressed, the American political system, disdain for aristocracy and social attitudes, and the vast westward expanses further removed the non-commissioned officer from his European counterparts and created a truly American NCO.

In 1778, during the long, hard winter at Valley Forge, Inspector General Friedrich Wilhelm von Steuben standardized NCO duties and responsibilities in his Regulations for the Order and Discipline of the Troops of the United States printed in 1779. Among other things this work, commonly called the Blue Book, set down the duties and responsibilities for non-commissioned officers and emphasized the importance of selecting quality soldiers for NCO positions. Von Steuben's regulations established the centerpiece for NCO duties that still remains today.

When all the Marine officers and staff NCOs were killed during the Battle of Chapultepec, it was NCOs who led the charge getting the Marines to the Halls of Montezuma. Sergeants and corporals have the same abilities today. It is the NCO who carries out the orders and makes sure that the Marines are capable of completing their mission.

NCOs do it all. These leaders have their hands in every aspect of the military, from the simplest daily activities to the most complicated strategic planning. NCOs are the first-line supervisors of the Marine Corps. From physical training and maintaining equipment, and from leading combat patrols to the varied tasks in between, NCOs handle all the daily activities throughout the Marine Corps.

There's a reason NCOs mold civilians into Marines. NCOs focus on training, mentoring, guiding, and leading. They are enlisted Marines taking care of other Marines and helping them solve their problems.

Carlton W. Kent, Sergeant Major of the Marine Corps from 2007 through 2011, said:

> *"The NCO is truly the backbone of the Marine Corps. I mean, look at the great NCOs over in combat. They're leading Marines and they're leading them with the legacy that we've always had in our Marine Corps. Our warfighting legacy is strong, as it always has been, and it's basically because of those young NCO's who are out there hooking and jabbing setting that great example for their young Marines."[7]*

Both General James F. Amos, Commandant of the Marine Corps, and Sergeant Major Kent continue to be committed to ensuring that Marines who wear corporal and sergeant chevrons have more responsibility, both in the rear and on the front.

When a junior Marine has a problem, it's the NCO who hears about it first. These NCOs are leading Marines in combat and can be the most senior Marine for miles.

They are creating the next generation of leaders.

They are also America's young men and women. Twenty-five percent of Marines are not old enough to legally consume alcohol, yet they're putting themselves in harm's way for something larger than themselves. The Marine Corps is easily the most junior in average age of all the military services.

And their job is more important today than it has ever been. Navy Admiral Mike Mullen, Chairman of the Joint Chiefs of Staff, told a group of non-commissioned officers that this is the most dangerous period he has seen in his more than forty years in uniform.

He said the threats of extremism and changes happening around the world associated with energy and resources make the present day "the most uncertain and potentially the most dangerous time since I've been serving." Mullen told a gathering of NCOs in Washington D.C. that their service at this time is absolutely vital. "[Your service] is bedrock to this country," he said. "Without that service and without that dedication, we could not be the country that we are; that's just flat-out the truth. We shouldn't take that for granted."[8]

Years ago, the United States Marine Corps fielded a recruiting slogan that promised potential recruits that the Corps builds men through body, mind and spirit. Over time and societal shifts, the words have been modified but the claim remains one of the most important bastions of the Marine Corps ethos. It has been expanded to include construction of women in body, mind and spirit in a Corps that must increasingly rely on females to fill crucial billets. Today, women comprise about 6.2 percent of the Corps. But the business of turning civilians—both male and female—into fully functional, utterly reliable, and highly motivated members of an elite Corps remains unchanged. And the annals of American military history indicate no outfit does it better than the Marines.

Why is that? Why is it that so many of our nation's pivotal leaders and most colorful characters credit service in the United States Marine Corps as a seminal experience in their climb to successful careers or the impact they have made on business, politics, arts, education and all realms of public service? Why is it that the common image of a tough, dedicated, capable and courageous fighting man always seems to involve a rock-jawed, steely-eyed Marine sergeant? There are as many answers to these questions as there are individuals involved. But the common denominator is that the Corps either gave them something they didn't have to begin with or shaped what they did have into valuable traits that lifted them above the struggling masses. Dig a little deeper into this question and another common quality appears: leadership.

If there's ever been a tough human quality to define, it is leadership. There's a simple dictionary definition (the capacity or ability to lead), and then there are countless varying viewpoints. Regardless of the differing opinions on what makes a good leader, leadership in the Marine Corps exemplifies unique characteristics. For instance, it requires fully committed leaders who are regularly tasked with making life-or-death decisions. That's likely why leadership skills learned or shaped by their military experience give veterans such a head start and unique perspective outside the strictures of service in uniform. Ask any number of Marine veterans and virtually all of them will say they learned leadership, or what things make both good and bad leaders, during their time in the Marine Corps.

That is because the Marine Corps treasures leadership above virtually all other qualities in the ranks; even above and beyond such obvious military virtues as bravery and tenacity in the face of danger or hardship. The Corps has a unique take on leadership that focuses downward rather than upward to the officers who are put in positions of responsibility as leaders of individual units. The Marine Corps believes with unswerving confidence that the strength of its collective body is in the backbone. And the backbone of the Marine Corps is its cadre of NCOs, especially the corporals and sergeants who push, pull and inspire when important things need to be done. There is in every Marine unit and in every formal school of its varied trades a zealous belief that real leadership reposes in the strength of its corporals and sergeants. That makes becoming an NCO a very big deal for young men and women in the ranks.

It also makes teaching leadership or honing the qualities that make a good leader one of the Marine Corps' most important pursuits. To be effective, leadership instruction must be adjusted to the individual and his or her situation, but that doesn't mean there are no yardsticks or common techniques that can be applied in the process of learning to be a leader. The Marine Corps runs leadership schools or learning programs for young NCOs at practically every

post, station, and command. The good news—especially
for Marines—is that it works. The bad news—especially for
everyone else—is that it's hard to get a grip on why it works
and how to quantify it beyond the obvious measures of
success or failure in the pursuit of common goals. But we
do know that young Marine NCOs are by and large
excellent leaders who get the most difficult jobs done in the
most difficult situations where failure often involves bloody
consequences.

We also know that effective leaders encompass certain
key qualities—the accepted term in the military study of
the subject is "leadership traits"—and that fourteen of them
are usually taught by acronym: JEDD J. BUCKLET II or
the more modern JJ DID TIE BUCKLE. This book
examines the leadership traits that make up that acronym:
initiative, bearing, unselfishness, dependability, endurance,
knowledge, judgment, enthusiasm, tact, decisiveness,
integrity, justice, loyalty and courage. One way to search for
answers to the leadership conundrum is by examining these
traits. In the pages that follow, we'll look at each trait
individually, examining what each one means based on
real-world experiences among young Marine NCOs. These
traits can be used as signposts on the journey to the understanding of what makes NCOs the backbone of the Corps.

In many of the chapters, there are revelations from
young men and women who have given no little thought to
the concept of leadership on the way to practicing it in the
real world. There are also studies of historical characters
long gone who have displayed NCO leadership to such
laudable degree that they've become icons to young leaders
in the modern Marine Corps. In keeping with the Marine

Corps practice of building leaders through body, mind, and spirit, the book is divided into three parts. The first shots downrange in this work are aimed at body, and as you'll see shortly, that involves a lot more than the physical fitness and stamina for which Marine NCOs are so justifiably renowned.

Table of Contents

Part 1

Body

Chapter 1

Initiative

If your ship doesn't come in, swim out to it!
—Corporal Jonathan Winters
United States Marine, 1943–46

The Marine Corps performs a variety of missions, some far beyond the usual amphibious landings and traditional combat campaigns. Now called "special operations," Marines rescue civilians from disasters, both natural and man-made. They board hostile ships much like in the days of the Barbary Pirates. They reinforce embassies throughout the world. In the ten years prior to the first Gulf War, the Marine Corps handled thirty-five of these kinds of emergencies.[9]

To hear Marine Corps NCOs like Sergeant Randy Burgess and Corporal Paul Spies tell it, lots of people see a job that needs doing or a problem that needs solving, but they just sit around complaining. That is just not in the DNA of these two men, who demonstrate the power of initiative and the Marine Corps' attitude toward that aspect of leadership.

Burgess had the thankless task of running a vital motor transport section when the Corps' 24th Marine Expeditionary Unit (MEU) landed in Somalia to help provide humanitarian assistance to a starving population and keep feuding factions from turning the country into chaos. It was stressful and often dangerous, even for accomplished mechanics like Sergeant Burgess.

In 1993, Burgess was stationed in a combat zone in Somalia, where he was responsible for all the vehicles and for keeping them moving. While in a convoy, Burgess got a call: one of the High Mobility Multipurpose Wheeled Vehicles (HMMWV, or "Humvee") was out of control. That was a serious problem. The Humvee was running at top speed and couldn't slow down. Burgess ran to the vehicle only to see it spinning as the driver tried to avoid hitting anything. The brake pads were wearing down before his eyes. Burgess took the initiative and jumped on the side of the Humvee, but he only had what was on his body to fix it, while it was spinning out of control at full speed. Burgess's maintenance and repair protocols never covered anything like this. He had to innovate.

Sergeant Randy Burgess (right) and Lieutenant Mark Shuster (left) at Fort A.P. Hill rest during a vehicle recovery with the Battalion Landing Team, 1st Battalion, 2nd Marines, Motor Transport Platoon, 24th MEU (SOC) September, 1992. (Photo courtesy of Mark Shuster.)

By fall 1992, roughly half a million Somalis lay dead from famine. Hundreds of thousands more were in danger of dying. The problems began in the 1980s when an insurgent group rebelled and proclaimed itself the Somaliland Republic. Tensions intensified as different rival factions proclaimed both Mohammed Ali Mahda and Mohammed Farah Aidid as the president. The resulting civil war, coupled with the worst African drought of the century, resulted in the loss of three hundred thousand lives. When clan violence interfered with international famine relief efforts, an American-led coalition was sent to Somalia to protect relief workers and the thousands of Somalis who were caught in the crossfire of a deadly civil war.

Operation *Restore Hope* was an American-led, United Nations-sanctioned unified task force (UNITAF) with authority to use all necessary means, including military force, to protect humanitarian assistance and peace-keeping operations. The coalition consisted of thirty thousand American military personnel and ten thousand personnel from allied nations.

On December 9, 1992, the U.S. Marines came ashore in Mogadishu and quickly established an expeditionary infrastructure to facilitate security and the delivery of food to the starving Somalis. On December 11, the Marines established a Civil Military Operations Center near the U.N.'s Humanitarian Operations Center. By doing this, the CMOC quickly became the national focus point for coordination of the military and humanitarian organizations.

The American military contingent covered an area of more than 21,000 square miles. Over these distances, units

conducted air assault operations, patrols, security operations, cordons and searches, and other combat operations in
support of humanitarian agencies. They also built or rebuilt
more than two thousand kilometers of roads, constructed
two Bailey Bridges (portable pre-fabricated truss bridges
requiring no special tools or heavy equipment for construction[10]), escorted hundreds of convoys, confiscated thousands of weapons, and provided communications. U.S.
Marines also participated in local civic action projects that
helped open schools, orphanages, hospitals, and local water
supplies. Due to these efforts, humanitarian agencies
declared an end to the food emergency, community elders
became empowered, and marketplaces were revitalized and
functioning.

Ultimately hundreds of thousands were saved from starvation, but unintended involvement in Somali civil strife
cost the lives of thirty American soldiers, four Marines, and
eight Air Force personnel and created an impression of
chaos and disaster.[11]

Up to their ears in the middle of the Somali chaos were
young Marine NCOs like Sergeant Randy Burgess. Burgess came from the Ozarks in southern Missouri, where
there wasn't a whole lot of opportunity for a young man.
Working on local farms, Randy learned to fix all kinds of
vehicles as the tractors had to run when the crops were due
to be harvested. There was no waiting for parts or for a
better mechanic—they just had to make it work. In 1988,
while Burgess was pumping gas at a local service station, a
Marine Corps recruiter started stopping by and soon
became a frequent customer. In response to the recruiter's

usual sales pitch, Burgess said, "Don't try to sell me. If I join, it'll be because I want to, not because you came in here and told me to."[12] The recruiter backed off and the two of them just talked.

Eventually, Burgess started asking questions. He wanted to work with his hands, and he wanted to see the world. He also stressed that he wanted to learn something that would translate to a civilian job after he left the military. Not much use for something like a 5974/Tactical Data Systems Administrator in the Ozarks. The recruiter turned to another page in the brochure: Diesel Mechanic. Sold.

Boot camp was tough, but Burgess was used to tough. He learned not to talk. His hillbilly twang came out low and slow, much to the delight of his drill instructors. As soon as he'd open his mouth, they'd start yelling "Faster! Faster!" For the first two weeks, he didn't speak at all if he could get away with it.

Major Mark Shuster, who was a Lieutenant when he worked with Burgess, shared his memories of their time in Somalia. "The interesting thing about Randy is that he was a very challenging NCO. He's from Missouri, he's a back-assward hick, but the man could fix anything that I asked him to." Shuster grinned fondly at the memory. "He had very rough people skills. Whether the Marines responded to him or not, I trusted him, to the point where if I needed something to get fixed and I gave it to him I knew it would get done."[13] Burgess regularly displayed admirable initiative.

Initiative means that when something needs to be done, you do it. You don't wait for orders or memos to tell you what you already know. It means staying alert and thinking ahead. It keeps you from being blind-sided by problems you didn't see coming. And it means using what you have on hand to attack those problems. You don't fail for lack of tools, you don't wait for just the right widget to resolve a situation, and you don't wait to see if anyone is looking so that you get the most credit you can.

Initiative means taking that first step. Not just any step—the first step—in something productive that has meaning. It takes a lot of creativity—even for just the simplest stuff.

Burgess remembered:

> *"We didn't have any paper for letters home ... Marines never have enough stuff. But that was okay. We would take MRE (Meals, Ready to Eat) boxes and cut out a piece of cardboard, write on one side, write 'Free' in the top right corner, and send it off in the mail ... and they made it home."[14]*

Members of any of the armed forces of the United States serving in a designated conflict zone can take advantage of the Free Mail system, where personal mail is sent as first class mail at no cost, as long as it is addressed correctly and has the word "free" written in the top right corner. Burgess's mother still has a few of these makeshift post cards from Somalia.

Initiative also played out for Burgess in more important ways while he was in Somalia. "As an NCO, you train up as well as down. Lieutenant Shuster was bitching about they should be fixing vehicles faster or something and I said, 'No way, sir. Can't treat my men like that.'" Burgess was pretty tough with the Lieutenant. His Marines were doing their best, and the challenges they were facing just weren't being appreciated. They were starting to resent the way they were being led, which caused frustration among the mechanics as the stress to get the job done increased. "Lieutenant Shuster told me he needed to walk away and think about what I said." When he did that, Burgess thought he was in trouble.

Shuster remembers the incident when he displayed some initiative of his own. He decided to work on vehicles when he could. Soon enough, he discovered he enjoyed it. Shuster found himself a jump suit, and he put some bars on the collar so that he would be recognizable to others coming into the facility. "So I went down one day, and I told Sergeant Burgess, 'Hey, I work for you today. Tell me what you want me to do, and I'll do it. I'm not your lieutenant right now, I'm just another mechanic.'"

Shuster learned a lesson by going down to work on the line. "If you run a company that makes widgets, if you're running the company, you're not expected to make the widgets. But it's very important that you understand how the widgets are made." It is vital that leaders understand what their teams go through on a daily basis. And it is a great way to connect with subordinates—the people who work for you.

"The best Marine Corps leaders say it all the time: The greatest asset, the most important asset we have in the Marine Corps are the Marines. It's not the trucks, it's not the rifles; it's the Marines. With what we do, with what we need to do, we don't need a rifle. You can kill a man with a helmet. You can kill him with a rock, or a spear,

*whatever it takes. Mission comes first, people are right
behind, but you've got to take care of your people."15*

By taking the initiative to work on the line with his
men—which is not a common Marine solution—Shuster
improved his ability to accomplish the mission while
discovering how to take better care of his men. The plan
worked. The relationships between all the Marines working
in the maintenance bays improved.

Shuster wasn't a mechanic, and he never claimed to be.
He didn't know that the men had to take off the entire
front end of a vehicle to change the water pump. In fact, he
wasn't sure what a water pump did. Therefore, he didn't
understand why the mechanics needed time to make the
necessary repairs. Joining his mechanics on the line helped
him lead them better:

> *"This novel approach worked because now, I was able to
> understand the maintenance requests, so that when I'm
> talking to my maintenance guys, and they're telling me,
> 'Hey, sir, that truck's going to be down for three weeks.'
> Why? Well, this is what it's going to take, and it's not as
> big a priority as this, and you start to understand your job
> better."16*

That was the real lesson. "Shared hardship is a great tool
for leadership. Share the work, be willing to get the grease
under your nails, and then, when it's time to be a Lieutenant, go do it," Shuster said. "So the impact with my troops
was great. And fifteen, sixteen years later, we still talk about
it. And it's nice. And it was Sergeant Burgess that set it in
motion."

In order for a team to bring innovative ideas to their leaders, trust needs to be in place first, so that they believe they will be listened to when they display initiative. In Somalia, trust had been built between the officers and the enlisted Marines. "That's a real key difference in my experience between the military and the civilian world," says Shuster. "There, people will tell you they're micromanaged to death, and there's not that sense of trust, because, my sense in the civilian world is that people are in it for money, and for themselves." But when people are working in a combat zone, it's such a different environment, and it requires a shared mission. Civilian companies function best with a shared mission, but the consequences of a poorly created or communicated mission is less than life-or-death. The Marines in the Motor Transport unit all had the same priorities: Get the job done, stay alive, be safe, and make sure to take care of the grunts, the guys who are out there every single day. Micromanaging is not part of the Marine picture. There's no time for it, and it hinders initiative.

Burgess didn't set out to take the initiative. He just wanted to come to work and fix trucks. He was fully committed to making sure that everything ran the way it should. And he was tough on the grunts. He was tough on his Marines. He wasn't perfect on paperwork, he wasn't perfect on procedures; he just got it done.

What is needed for innovation to flourish for small teams? Support and protection of leadership, access to resources, autonomy, ownership, and the permission to fail.[17] Although often problematic, the Marine Corps more often than not provides just this kind of environment to its Marines, especially in combat.

Because of the uncertainty, disorder, and fluidity of combat, the Marine Corps pushes significant decision-making authority down through the ranks. With an understanding of the mission—the commander's intent—those closest to the action can take advantage of ground-level information not readily available to their superiors.[18] When NCOs have the freedom to take the initiative, individuals can identify an opportunity, take action, and lead others to exploit the objective.

In the words of General Charles C. Krulak, who served as 31st Commandant of the Marine Corps:

> *"The inescapable lesson of Somalia and of other recent operations, whether humanitarian assistance, peace-keeping, or traditional warfighting, is that their outcome may hinge on decisions made by small unit leaders, and by actions taken at the lowest level."*

Sergeant Burgess is used to making decisions and taking action. Once, he was faced with a broken vehicle battery, and no replacements were to be found in Somalia. Fortunately, he never threw anything away. He found some pieces of lead that weren't useful for much anymore, put them in his canteen cup, and melted them down. Then he taped some cardboard to a tube, made a mold on the battery, poured the molten lead, and waited for it to set. Once it had cooled, he had a working battery. He recalled how a little initiative went a long way:

> *"Now, this is not the accepted procedure for battery repair. But the accepted procedure would have meant another*

truck down which was not acceptable to me. People in the civilian world ... as soon as they say 'policies and procedures' I know they can't or won't make a decision, and I turn off."[19]

Burgess led in the same way he repaired vehicles. Joe Ford served as one of Burgess's Marines. Serving under Burgess gave him a unique perspective of his sergeant. When Burgess first came to Ford's Motor Transport unit, Ford was a Marine Integrated Maintenance Management System (MIMMS) clerk. That job included updating the status of all equipment for the battalion, ordering parts, and tracking their status. Burgess took over the shop chief position soon after arriving, putting him and Ford together working hand-in-hand every day.

Burgess taught Ford how to stand his ground and argue a point when he believed in it, no matter the consequences:

"If Randy knew he was right, he wouldn't give up, especially when it came to his men. A lot of people like the fact that they are 'in charge' and like to make that fact known. Randy didn't. He would rather everyone worked in unison until the mission was accomplished."[20]

Marine legend has it that STEAL stands for Strategically Taking Equipment to Another Location. It is often a way of life and a key to survival in the Marine Corps, which typically never has enough stuff or the right stuff. There is rarely malice involved, but some Marines "tactically acquire" gear that they need. Rumor has it that there was one thief in the Marine Corps 225 years ago, and everyone else has been trying to get their stuff back ever since.

In 2006, only about 15 percent of the total Marine Corps personnel were deployed to Iraq at any one time, but since then the Corps has deployed about 40 percent of its ground equipment, 50 percent of its communications equipment, and 20 percent of its aircraft to Iraq. Yet, according to a 2005 report by the Marine Corps Inspector General, the Marines in Iraq "don't have enough weapons, communication gear, or properly outfitted vehicles."[21]

One reason for the systemic lack of resources comes from the Marine Corps' unique role within the Department of Defense. Compared to the Army, Navy, and the Air Force, the Corps is the smallest in size and resources. The Marine Corps active personnel, as of July 2010, consists of about 625,000 Marines compared to the Army's 562,400.[22]

The Corps' budget also is the smallest of the armed forces. Looking back to fiscal year 2007, the Marines requested about $18 billion, while the Army budget amounted to about $110 billion. Even on a per capita basis, the budget for land forces in the Marines is 35 percent less than the Army. Some of this is due to the fact that unlike the Army, the Marines don't have a separate military department, and therefore must also compete for funding with the Department of the Navy. During fiscal year 2007, the Corps received 14 percent—or $18 billion—of the Navy's $129 billion budget.[23] Since the allocation of the budget to the three military departments has been relatively fixed historically, any increase in the Marine Corps budget generally comes at the expense of the Navy, not of the Army or the Air Force.

Another reason for the Marines' traditional lack of supplies in combat is that they are often placed well ahead of supply lines due to the nature of the type of operations they conduct. And so, Marines with initiative find ways to get what they need—even if it means going outside traditionally approved supply channels.

Sergeant Burgess faced this problem in Somalia. Not being able to get the supplies he needed to repair the vehicles under his care, he did what any good Marine does: he went over to the Air Force supply dump with a few of his Marines, list in hand. When an officer discovered them looking around at their inventory, Burgess was concerned—but the Air Force was on their side that day. The officer asked for the list of parts and helped them gather what they needed.

Initiative and creativity often are needed to get the job done, although they're not always rewarded and often go against regulations. But not everyone wants to fight regulations, and some are afraid to do things that aren't by the book. But, Burgess notes:

> *"NCOs can help develop initiative if the guys know you have their backs and that finding a solution won't put them in jeopardy. And it's necessary to do so in the Marine Corps as there's never enough of the 'correct' stuff. Innovate, adapt, overcome."*

Burgess says that when system works, it's excellent for training leaders. When it doesn't, the best people leave. It can be a real retention problem—the ones staying in are the ones who do the minimum, while the best leave when they feel unappreciated. Motor Transport and mechanics have it tough. It's not a glamorous job; it's hard work, with not much time off and a lot of stress and sweat. It's a challenge to keep the Marines motivated.

Joe Ford, the Marine who served under Burgess, was motivated. Today, he is vice president of a trucking company in Dayton, Ohio, that has three companies under its flagship. Ford runs all three companies including the shop. It is set up exactly how Burgess set up their Marine Corps shop. Ford has seventy-three employees who count on him and his decisions every day. Burgess taught him how to challenge the standard operating procedure if there is a better way. He also taught him that if he didn't like something and had a better way to do it, that it was in his own power to change it.

Burgess showed Ford that, by taking the initiative, he could grow professionally and he could become a great leader. For example, on their way to Somalia, the company's MIMMS clerk before Ford had made some major mistakes. Burgess asked the lieutenant to recommend Ford for the job. Ford took the job even though he thought it would stretch his limits:

> *"I was scared shitless as this was a huge responsibility, but I did it and did it well. He knew I could do it when I didn't know myself. Burgess taught me how to be a leader; he taught me how to be a man. He taught me how to be a leader of men."*

Once Sergeant Burgess and Lieutenant Shuster left Mogadishu, they spent most of their time in the south, in the city of Kismayo.

Kismayo is Somalia's third largest city near the mouth of the Jubba River, where it flows into the Indian Ocean. An important port city, its estimated seventy thousand inhabitants were in the midst of constant clan conflict. Formerly one of the Bajuni Islands, the peninsula subsequently was connected by a narrow causeway when the port was built, with United States assistance, in 1964. The port served as a base for the Somali Navy as well as the Soviet Navy after a military coup in Somalia in 1969.[24] Shuster and his team were able to participate in some weapons searches, but his primary role was to provide support to the grunts who were stationed at different outposts across the area of operations. Shuster's team was completely engaged with the local populace, the warring factions, and the tribal leaders. The Marines of this unit were fixing trucks, going on convoys, and making sure the Battalion Commander and the rest of the Marines could get what they needed. Shuster spent his time making sure his unit was functioning correctly, that the missions were being accomplished, and that his men had the training they needed.

Training is key to Marines, and it doesn't end when boot camp is finished. Marine leaders are forced to make life-or-death decisions in the midst of chaos. Training readies them for that. It also prepares them to take the initiative and make decisions on the ground, regardless of whether things are going well. Burgess notes that:

> *"You're going to get exhausted, be confused, get yelled at in a firefight so you need to train that way so you don't stop in combat. Train like you fight, fight like you train. Must move forward always; the worst mistake is stopping. You'll never have enough information. But you can't let that stop you. Make a decision and act on it, but keep open to new information so you can adjust as you go."*

The common thread uniting all training activities is an emphasis on fostering integrity, courage, decisiveness, mental agility, personal accountability and, of course, initiative. These qualities and attributes are fundamental and must be aggressively cultivated within all Marines from the first day of their enlistment to the last.[25]

While on patrol in downtown Mogadishu with Weapons Company, 1st Battalion 2nd Marines, Burgess received the call about the out-of-control Humvee. The accelerator pedal on the Comm vehicle was stuck and the engine was running wide open. They were burning up the brakes trying not to smash in to the vehicle in front of them. Randy knew exactly what the problem was; it was not the first time he had heard this complaint about a Humvee. He also knew it would take all of his initiative to make a repair on the fly. He got on the radio and told the Company Commander he needed stop the convoy for repairs. The commander wasn't too happy about this, as they were not in the nicest neighborhood in town.

But the only other option was not good. The vehicle contained sensitive equipment, so if it was deemed inoperable and no tow truck was available, it would have to be rendered useless to the enemy. That meant removing as much equipment as possible and destroying the rest. In other words: blow it up. Burgess let the company commander know he would be making note of his refusal to stop, so he could report to the Battalion C.O. why he had to blow a Humvee on a routine patrol.

Burgess was more than a mouthy corporal. He was the only Marine on the patrol with his particular military occupational specialty. That made him the acting Maintenance Chief, a billet normally filled by a Staff NCO, as well as the Motor Transport Officer, generally a Lieutenant's billet.

While deployed, the Marines were using JP5 fuel for the Humvees, instead of the preferred diesel fuel. JP5 is used on ships and helicopters, and since storage on a ship is limited, having to carry only one kind of fuel was preferable. "Humvees were equipped with a 6.2 litter GM diesel engine which would run on JP5 and Kerosene," Burgess explained:

> *"but it didn't mean that it was good to do so—a fact that I could never get through anybody's head. The fuel injection pumps on the Humvee had no source of lubrication for its internal parts, other than from the diesel fuel. JP5 isn't as 'oily' a fuel, so it doesn't provide the necessary lubrication, and this causes the seals in the pump to dry out."*

One of the main seals affected by this is the seal that is around the accelerator lever rod, where it goes into the pump. The addition of salt air and salt water from beach landings makes the problem even worse, and the accelerator lever return spring becomes rusty and weak. That made for a deadly combination of dry rubber bushings and a weak return spring, resulting in a sticky accelerator pedal.

If they had been in the rear, it would have been possible to add an extra return spring, or get a stronger spring, keep the Humvee running as long as possible, and then eventually evacuate the vehicle to higher maintenance to get a new pump installed. It was very common to have to do this for more than half of the fleet when returning from a deployment on ship. However, when in the field, other options are required. Having spent more time in the field than in the shop, Burgess excelled at coming up with other options. His years of keeping tractors and other farm equipment running as a farm hand back in Missouri also helped.

After convincing the C.O. to stop, the convoy finally pulled over to a fairly safe position. Several Marines were put in place in a defensive circle around the vehicle. Burgess maneuvered his vehicle as close as he could on the narrow streets:

> *"and ran up to the vehicle armed only with my M16 and my needle-nose vice-grip pliers. I quickly tried to reattach the spring to a different location on the engine to give it more pull, but the pedal was still sticking. I was getting several orders from unknown voices to hurry the hell up."*

The streets were beginning to fill with angry-looking Somalis, and Burgess and his fellow Marines knew they were a choice target. Despite the threats, Burgess set to work. Right away he had to improvise, using a part of his uniform to fix the problem: his "boot band."

Marines don't tuck their trousers into their boots (like they do in the Army). Instead, Marines use a device called a boot band, which basically is a giant cloth-covered rubber band that goes around the top of the boot; the bottom of the trousers are tucked under it. The typical boot band doesn't last very long, so Burgess chose to wear a different kind of boot band, one that resembled the spring of an old screen door. Most Marines didn't like this type of boot band as it is tight, and often digs so hard into boot that it leaves grooves in their legs. However, the longer life expectancy of these boot bands has lent them the reputation of being "the only things left with the cockroaches when the earth is destroyed."

Luckily for the convoy, Burgess was wearing his "screen door boot bands" that day in Mogadishu, and with that in hand, he was able to improvise a solution:

"I took of one of them and hooked it to the accelerator lever on the fuel injection pump and stretched it to the front of the engine where I could hook it, just behind the power steering pump. I yelled at the driver to push the pedal, he said it was better, but I could see it wasn't returning all of the way and the engine was still running too fast. I needed to stretch the boot band a little farther, but there wasn't anything else to hook it to."

Burgess could hear the Somalis making a bunch of noise nearby, while the Marines were yelling at him to make a decision. He had to act fast:

"I quickly took a boot lace out of one of my boots, tied it to the free end of the boot band, and stretched the boot lace over the front of the radiator and down to the radiator bracket. Using a trucker's knot, I pulled it all tight, stretching the boot band just enough to keep the accelerator from sticking. The driver gave me the thumbs up and we got everyone loaded into the vehicles and got the hell out of there."[26]

Burgess was glad to get back to base and get out of his vehicle—until the company gunnery sergeant saw his uniform without his boot bands. No sooner did Burgess and his fellow Marines get back to base camp when the gunny yelled at him to blouse his boots.

Burgess routinely displayed that kind of initiative. Shuster recalled how his fellow Marines knew they could count on him:

> *"Burgess was sent out because we knew that if anything broke, he'd be the guy there. The Battalion Commander told me, 'Burgess. That guy is amazing. He can fix anything; he deserves a Bronze Star for this.' He got nothing, though. It became one of those folklore things. The guys still talk about it today. I really didn't look more into it even though it seems so unbelievable because I felt, 'Yep, that's Randy.'*
>
> *Today, people would have had flips, cell phones, would have recorded it and it would have been all over You-Tube. Back then, though, we didn't have that stuff with us."* [27]

But Burgess hadn't acted with initiative because he hoped he would get a medal. He did what he needed to do to get his fellow Marines back to safety. The situation he faced in Somalia demonstrates a sort of hands-on, get-the-job done at all costs initiative that is a hallmark of good Marine NCOs. But there are other examples in which relatively junior Marines can and do think on a larger scale to solve problems with bigger pictures.

Take the example of Corporal Paul Spies, who was thinking about road trips he'd taken along American highways as he ran patrols dodging improvised explosive devices in Iraq in 2010. He recalled signposts on practically every stateside roadway noting that one stretch of road or another was kept clean and litter-free by various sponsoring organizations in surrounding communities. He wondered whether something like that work on the mean streets and hazardous roads of Iraq.

Corporal Spies was a combat engineer with 9th Engineer Support Battalion, 1st Marine Logistics Group (Forward). His idea: The "Afghan Adopt a Road Initiative." The plan: for every thirty days Afghan villagers help keep their roads clear of IEDs, they receive aid-based incentives in return.[28] Now it was his turn to brief unit commanders of Regional Command, South West during a conference on countering IEDs.

He explained the plan, outlined in a fifteen-page proposal, and the possible incentives. Medical care could be given to the villagers if no IED incidents occurred in a given month. Points could be awarded monthly, and the points could be used to fund projects such as schools, wells, and irrigation.

Spies developed his plan and shared it with his immediate superior, who passed it to his boss. When the company commander saw it, he was impressed, and it continued up the chain of command. Eventually, it hit the desk of Major General Richard P. Mills, who loved the idea. This led to the invitation for Corporal Spies to brief commanders at Camp Leatherneck.

His battalion commander, Lieutenant Colonel Ted Adams, commanding officer of 9th ESB, agreed that it wasn't necessarily technology that would win against the counterinsurgency that characterized the fighting in Iraq:

"I am always challenging Marines to come up with a smarter way of doing business. I know the way we'll be more successful in this fight is in our ingenuity. Technology isn't always the answer, smart Marines are."[29]

Like Burgess, Spies found a way to take the initiative. He didn't do it for recognition. He didn't do it for glory. He did it to help his fellow Marines. Burgess and Spies both knew that they were entrusted to think outside the box, to try new things, and to consider new solutions to existing problems. Those who led them encourage that kind of thinking, that kind of initiative. And, in turn, that kind of initiative is what turns NCOs into leaders.

There is no question that smart NCOs like Sergeant Randy Burgess display outstanding initiative. But they also demonstrate bearing, which is another core leadership trait of the Marines.

Chapter 2

Bearing

"Attitude is a little thing that makes a big difference."
—Winston Churchill (1874-1965)

For navigators or aviators, the term "bearing" means a direction or heading. For Marine NCOs, the word has a different connotation exemplified by people like Donald MacGregor.

Sergeant Donald MacGregor is a man with bearing. He doesn't as much meet you as he confronts you. His eyes twinkle. His hands shake a bit due to recently-diagnosed Parkinson's disease, but that doesn't bother him and it certainly doesn't lessen his impact on those who can get him to talk about his days as a U.S. Marine. He still has a powerful presence. "I write a little slower than before, that's all. Of course, I do everything slower these days," he said with a chuckle.

MacGregor's youthful attitude and energy belie the fact that he's been alive for about a third of the entire history of the Marine Corps. The many fantastic events of which he has been a part—filled with extraordinary men, softened around the edges by the lives of glorious women—live in his attitude as much as his memory. His bearing today reflects a distinguished past.

MacGregor was one of the fighting Marines in the Pacific during World War II. He was a tough, courageous man in those days, and he still looks the part. The man has bearing in spades, but that's a difficult quality to define.

In military circles bearing has more than one definition, all of which are closely related. Bearing is generally credited to someone who looks good in uniform; who has a certain presence, charisma, or aura; one who shows grace and poise through posture and gestures. But bearing goes well beyond surface appearances. Other definitions point to direction or position, as in "the pilot headed toward a new bearing." That sense of the word points to someone who has a sense of direction: a goal or a purpose guides his behavior. This sense of movement toward a goal and the ability to inspire others to follow along the path is the gut-level sense of the word. The leader who seems to be heading somewhere important and interesting has more followers than the person wandering aimlessly. That aimless Marine may have a goal, but without a sense of direction to travel in reaching that goal, he lacks the ability to inspire others to follow their path. He lacks bearing.

The Marine with bearing is driven. An NCO with bearing is driven toward a goal with purpose, jumping at opportunities for self-improvement that increase his ability to reach that goal. These leaders know their own strengths and weaknesses. They know which opportunities will be most important and which will help them be better leaders. Bearing is also about channeling that drive to other people. Leaders who channel stress and anxiety actually repel other people. Leaders who channel a sense of determination and purpose will attract people who want to emulate that dynamic behavior. That's where bearing pays huge dividends in effective leadership.

Bearing also involves having a deep awareness of situational and human environments. NCOs with effective bearing know where they stand and they understand the environment in which they work. They know the Marines around them and are sensitive to their needs. They have a sense of humor, when appropriate, to reduce stress levels when a situation requires that. They see what needs to be done, and they do it in the best manner possible, thus setting an example for others to follow in both attitude and behavior. Sergeant Donald MacGregor's experiences in World War II taught him about bearing—through a very tough education.

MacGregor's path into the Marine Corps was rather twisted. He was driven to excel even as a young man and had hopes of becoming an officer. While in high school, he was in the Reserve Officers Training Corps for three years, ending up with the rank of cadet captain. He apprenticed as a machinist and became a journeyman at an unusually young age. He even completed a couple of years of chiropractic college in his spare time. When he decided to join the military in 1943 to help the war effort, he expected to join the Army and apply what he had learned from ROTC. But that didn't happen.

When he arrived at the enlistment center in Los Angeles, he was told to take off all his clothes except his socks and stand in line. In 1943, enlistment levels were high due to increasing military manpower requirements being met both by volunteers and by those being processed through the draft, so lines were long. It was a weird experience to stand among a gaggle of mostly naked men, and MacGregor was beginning to wonder if he was making a wise choice. There were curious blanks in the line: a group of guys, then no one, then another line of guys, then no one in front of them.

As he progressed through the line, he understood the reason for its staccato rhythm. The line moved next to a wall, and at regular intervals, knee-high windows provided excellent viewing of the men without trousers to the staff of the offices on the other side of the glass. The viewers were delighted with the show. The men in line, however, were not enjoying their immodest display. So the line would pause right before the windows, and when there was room on the other side, they would run as fast as they could past their civilian audience.

After receiving his paperwork, MacGregor endured the usual routine of "turn your head to the right and cough—bend over grab hold of your ankles—and smile." Finally, he was allowed to get dressed, take his paperwork, and head over to where the Army was processing enlistees.

Along the way, a Navy chief petty officer sitting at a desk asked MacGregor for his paperwork, which he passed along. At this point, MacGregor had his introduction to military bearing. Looking over MacGregor's papers, the chief said, "We need you in the Navy. You got a machinist background, and we need machinists." MacGregor wavered, still thinking of a possible Army commission. But the chief was persuasive. With very few words exchanged, MacGregor found himself following the chief's guidance and moving through the double doors to where the other Navy recruits were waiting.

MacGregor was not sure what they would have done had he refused to go. By following the chief's will, he was on his way to becoming a Seabee in one of the Navy's Construction Battalions.

Seabees were a priority, so candidates were processed more quickly than the other recruits. But then MacGregor got another switch. The Navy discovered that through his chiropractic education, he had studied biology, physiology, toxicology, and diagnostic medicine. Now, he was going to be a corpsman. Hospital corpsmen serve as enlisted medical specialists serving the U.S. Navy and the U.S. Marine Corps, a job similar to the Army's medics.

MacGregor recalled what happened next:

> *"They shipped us out to Quonset Point, Rhode Island. None of us had ever heard of it. We got six weeks training. Period. They put us through battle dressings, wounds; everything was included during those six weeks, including their obstacle course."*[30]

The Marine Corps uses obstacle courses to build the confidence necessary to develop bearing. These courses also develop the ability to think and remain focused while under pressure. MacGregor can still remember the pressure:

> *"And this was dead of winter. Part of the course included one of those metal-link bridges. That bridge was impressive to me because it didn't have any railings on the side. It had rained and turned icy, and we had to run across it. I made it; several guys didn't—but you didn't fall but about fifteen feet. At least you fell onto an ice pond."*

After training, the young sailors were put aboard a train and told they were heading to Florida. All the windows were covered. Then the train turned and they found themselves headed for California. Their final stop: Port

Hueneme, where MacGregor got hit with another switch in his military fortunes. First the Army, then the Navy: now MacGregor found himself in the Marine Corps.

The 4th Marine Division was formed on August 14, 1943, by reorganizing and shuffling other units. But they still had a need for additional manpower, and were gathering Marines where they could. The weapons company of the 23rd Marine Regiment, part of the 4th Marine Division, was preparing for war at Port Hueneme—and now, so was MacGregor. His bearing faced a new challenge while he did his best to adapt to his new role and train for war.

All the instructors at Port Hueneme were Marine veterans of the 1942 battle of Guadalcanal, and they made the training intense. MacGregor recalled that "They taught us Judo, and they were determined that you were going to feel it." He appreciated this training and saw how it was helping him to develop. MacGregor had never been a large man. Now, however, he was starting to see how his physical presence was changing the way he was perceived. He could tell that joining the Marine Corps opened up a whole new set of goals and opportunities for him.

MacGregor and his fellow Marines soon were shipped to Pearl Harbor on Oahu, Hawaii, where he was pressed into unofficial medical service for his unit. "They still had me doing corpsman duties, such as passing out APC pills," he recalled. APC pills, often referred to as "Aches, Pains, and Complaints" or "All-Purpose Capsules," were a combination of aspirin, phenacetin, and caffeine. MacGregor had learned that Navy corpsmen used these medications as a standard pain reliever and fever reducer and he distributed a lot of them to his fellow Marines when he wasn't busy with his less pleasant medical chores.

"I didn't mind most of it," MacGregor said, remembering the routine. "But I got real tired of the short-arm inspections." These medical examinations for symptoms of venereal disease were frequent and tiresome—especially for the man performing them over and over again. MacGregor wanted to stay on his new path and become a leader of Marines rather than staying stuck in the medical support role. So when an opening came up for an opportunity to train to be a hard-hat deep-sea diver, MacGregor jumped at the chance.

Sergeant Donald MacGregor in Hawaii, 1943, attended dive school on the island of Maui before heading into battle in the Pacific Islands during World War II. (Photo courtesy of Donald MacGregor.)

Divers were of great use during the war in the Pacific, where ship damage and malfunction happened in places where there were no dry docks for repairs. In 1945, an official diving school was established at Marine Corps Air Station Cherry Point in North Carolina. The physical requirements for diving at that time were so stringent that only eleven out of one hundred twenty applicants were accepted for the course.[31] MacGregor had the physical ability and energetic attitude necessary to qualify for a less-formal course of instruction offered in Hawaii. He saw the training as a chance to improve himself, learn a new skill, improve his physical conditioning, and be better prepared to handle the Pacific environment in which he would be fighting and leading other Marines in combat.

Few institutions are as attentive to the intertwining elements of bearing as the U.S. Marine Corps. A unique and elite fighting Corps, its history demonstrates its past commitment to high standards, and it continues to demand excellent bearing. Marines are expected to stand tall, and to always display pride in their uniform.[32] The Marine Corps expects that you look like what you are: a Marine.

After he qualified as a diver at the beginning of 1944, MacGregor would have his chance to test his bearing under combat conditions. He and his fellow Marines were taken aboard a ship on Oahu to rejoin his outfit as the 4th Marine Division was leaving Hawaii and heading off to war. He joined them in the attack at Roi-Namur in the Marshall Islands, in the northern part of the Kwajalein Atoll. From January 31 through February 3, 1944, the American commanders took the lessons they had learned from the previous battle on Tarawa in the Gilberts and launched twin assaults in both the north and the south during the campaign in the Marshall Islands. Outnumbered and under-prepared, the Japanese defenders resisted as best they could, but only fifty-one survived of an original garrison of 3,500.[33] The 4th Division returned to Maui at the end of February. MacGregor survived and had a little time to contemplate the bearing he had observed among Marines in combat. He wasn't the only one who appreciated that.

On April 26, 1944, Admiral Chester Nimitz journeyed to Camp Maui to present awards to men who had earned them in combat at Roi-Namur, and read a special commendation for the unit: "The world knows of the gallant performance and achievement of the men who fought at Roi and Namur Islands ... There, the 4th Division wrote another brilliant chapter in the chronicles of the Marine Corps."[34]

On May 29, about three months after returning from Roi-Namur in the Marshall Islands, the 4th Division sailed for Saipan, the tactical center of the Marianas Islands, and MacGregor was back in combat, where Marines were deployed alongside Army units. During the assault, he saw a problem developing on the beach. "I noticed a difference between the Marines and the Army in combat," he said:

> *"If you've got an invasion and you're going in on that invasion, and you come to a pillbox, the Marines hit that pillbox, but then they go around it and keep going. They know the pillbox isn't the goal, but an obstacle on the way to the goal."*

This concept is a key to bearing in leadership. The leader with bearing understands the destination, and keeps moving forward. Without that orientation, when a person hits an obstacle on the way to their goal, overcoming that block can become a goal in and of itself. When that happens, the leader loses sight of the greater mission, which often robs vital momentum. MacGregor noticed that the Army and the Marines approached obstacles in a different way:

"On Saipan, when the Army came in, and hit a pillbox, they stopped the entire line of their troops until they knocked out that pillbox completely. Then they went ahead. This kind of left the Marines here and the Army there, you know? So we had infiltration on both sides."[35]

In the midst of combat, MacGregor was showing his situational awareness, a key element in the bearing he was developing as a Marine and as a leader.

This awareness includes an appreciation of what is happening in the environment that will affect any potential goal. Many people have little or no situational awareness. They go through life without noticing much of what is around them. Leaders with bearing understand what is happening around them, and they use that information to anticipate future events.

After a period of rest and retraining on Maui, MacGregor was back aboard ship with his unit, heading for another objective in the Pacific. Listening to the radio, they heard Tokyo Rose, which was a nickname given to any of about a dozen different English-speaking female broadcasters specializing in Japanese propaganda. It was she who gave them the first hint of where they were heading. "We know the 4th Marine Division is going to Iwo Jima—I bet you guys don't even know that." And she was right. The Marines did not know their destination. MacGregor recalled that he and the other men of the 4th fell silent upon hearing that Iwo Jima was their destination. None of them had ever heard of it, much less knew where it was. But he and his fellow Marines were about to find out. MacGregor and the other Marines were told it would be a seventy-two hour campaign, on an island that was only eight square miles. How hard could it be? They would shortly find out.

Iwo Jima was vital to progress toward the end of the war in the Pacific. Only 758 miles from Tokyo (by comparison Hawaii was almost four thousand miles away), Allied forces in the Pacific needed the airstrips for planes to be able to refuel before hitting the Japanese homeland. And they needed to deny the Japanese the use of the air fields for their planes.

Although the United States had prepared the island for invasion during months of Navy gunfire and heavy air bombardment, the Japanese were equally prepared and undaunted. Enemy formations on the island included many artillery and antitank units. Food and ammunition had been stockpiled. The Japanese had built bunkers and blockhouses, and perhaps most importantly, had created a huge network of tunnels and caves that our bombs had barely touched.[36]

Sergeant MacGregor went in with the second wave. Before they climbed down the cargo nets from the mother ship to the LCVP (landing craft, vehicle, personnel), he loosened his gear in case he landed in the water:

"In the water, there were swells. They didn't bother the mother ship very much, but the small landing craft into which we were headed were going up and down with the swells. They get as close as they can, you know, but they don't have anything to hook onto, and that would have torn the LCVP apart if they had hooked up. I loosened up my cartridge belt in case I went overboard. I didn't want all the weight of my stuff to pull me down. They had only

given me one magazine anyway; fifteen rounds for my
Browning Automatic Rifle."

Once on the beach, MacGregor reassembled his gear
while he assessed the condition of the battlefield. The sand
was like marbles. Marines would slide down one foot for
every two they went up. Even by crawling, they were able
to gain little traction. Surrounded by chaos, MacGregor
kept his bearing and prioritized what he needed to move
forward effectively. He spotted a man carrying two cases of
ammunition and went to lighten his load. Fifteen rounds
goes through an automatic rifle awfully quickly, and
MacGregor knew he would need a lot more ammunition
for his weapon. MacGregor helped the Marine carry the
cases, following him up terraces that were scattered along
the beach, some of which were five or ten feet tall.

MacGregor never learned the man's name. He was hit
in the chest with an anti-boat round, about the size of a
40mm shell.

"It exploded and I remember I was looking at him when I
saw his ... saw it blow him apart. He took a step and fell
into the sand. And I thought to myself, 'He's going to
suffocate.' How ridiculous is that? I just saw him blown
apart; he's not going to breathe. But you have to just keep
going."

MacGregor did. He took that ammunition, resupplied
himself and his Marines, and fought on that beach for
another four days. On that day, MacGregor's bearing was
tested even further:

"I think it was D+4, the day before they raised the flag.
A guy came along the beach and yelled, 'We need a diver!

We need a diver!' 'Hell,' I thought. 'I'm a diver, and here I am in this slaughterhouse,' you know, and it was a slaughterhouse."[37]

The man was desperate. A ship was disabled, and if no one could be found to repair the problem, urgently needed supplies would be kept at sea, away from the Marines who needed them on the beach. Japanese guns had made it extremely difficult to land supplies, destroying dumps that contained much of the initial resupply of ammunition and demolitions.[38] Every landing craft that still could transport supplies had to be kept functional to land gear and evacuate the wounded off of the beach to hospital ships.

MacGregor stepped up and volunteered for the dangerous duty of diving into the water to make the repairs on the crippled vessel while under fire. He was taken out to the disabled landing craft. Up on deck, he met the captain, who explained the situation. The landing craft had tied up alongside a big transport ship filled with critical supplies to be unloaded. Unfortunately, their hawser, a thick cable for mooring ships together, had gotten tangled in the screws of the landing craft. Someone had to go over the side, under the vessel and cut it free.

MacGregor asked for diving gear. The captain brought out a mask—and nothing more. No other equipment was available, not even a suit. The mask was a full-face rig that allowed the diver to breathe underwater, with air supplied by a two-man pump on the ship that looked like a piece of railroad gear.

MacGregor went over the side. He could see that a thick hawser was the problem. He was given a hacksaw to cut through the hawser and untangle it from the screws. The saw was barely sufficient for the job. MacGregor reminded the crew to keep the engines turned off. He would be working next to the large screws that provided propulsion and direction to the craft. If they were to turn, which they could if the engines were running, MacGregor would be in grave danger: "Those big screws are right there. I get down there, I've got my legs wrapped around the propeller shaft, and I start sawing on the hawser. And the boat starts to move." When the shaft began to turn under his legs, MacGregor kept his bearing and didn't panic. He kicked himself loose from the shaft and swam up to the surface as quickly as he could. Fortunately, his own lines hadn't fouled so he did not lose his air supply. The Marine who had been on watch knew to keep the engine shut down, but signals were crossed and the new man on duty hadn't gotten the message.

Ordinarily, when these craft were in port, the propeller is kept turning slightly to avoid leaks around the engine. Keeping his cool under pressure, MacGregor was able to halt the engines, impress upon the crew the importance of listening to his orders, and complete the repairs. Maintaining his bearing was a critical part of accomplishing the mission and reestablishing a vital link between the invasion fleet and the Marines struggling on the bloody beaches of Iwo Jima.

People are drawn to leaders like Sergeant MacGregor, who kept his focus on the ultimate goal. And it is human nature to try and be like the people we admire. The bearing that develops from that focus is also a way for leaders to influence others. If a leader can be the kind of person others admire, he or she can guide their behavior, standards, and priorities. And if leadership is the art of getting things done through other people, bearing becomes a powerful tool. Use it cautiously and wisely; infamous tyrants and dictators almost always have remarkable bearing. For this reason, the Marine Corps doesn't rely on bearing as an independent gauge of a great NCO. The Corps also expects leaders to have other valuable traits, such as unselfishness.

Chapter 3

Unselfishness

We, the officers and NCOs, owe it to the men we command and to our country that we make ourselves fit to lead the best soldiers in the world, that in peace the training we give them is practical, alive and purposeful, and that in war our leadership is wise, resolute and unselfish.
—Field Marshal Viscount Slim of Burma, 1949

It's tough to remain unruffled, to maintain bearing, under stress. When that stress involves time in captivity, Marine training can be the difference between survival and sinking into the depths of despair.

Embedded within the Marine Corps culture are its two main priorities for all leaders, no matter what the unit size: accomplish the mission and then take care of your Marines. Generals know this; corporals whisper it in their sleep. The first priority, to accomplish the mission, is easy to comprehend. When tasked with a job, that job needs to be done, whatever it takes. For most people, that does not mean putting others at physical risk to fulfill that obligation. Marine leaders, however, can—and often do—find themselves in situations in which lives are on the line, and that makes for some difficult decisions. But the mission comes first.

The second priority also seems clear: Take care of your Marines. Leaders need to care for those they lead. Good leaders make certain that their people have the supplies they need, that they understand their specific jobs and the overriding mission guiding them, and that they have enough time away from the job to handle personal business and to get sufficient rest.

Unselfish leaders make decisions that benefit as many as possible, without worrying too much about themselves. They look out for the welfare of their teams beyond simple job descriptions, legal concerns, and even their own personal comfort. And they do this most particularly in difficult circumstances.

Good Marines of all ranks and in a variety of circumstances demonstrate unselfishness every day. There are examples throughout the NCO corps: the sergeant who makes certain that everyone in his unit eats before he does; the corporal who shares her water when it is scarce and encourages others to do the same; the leader who gives credit to subordinates and attempts to get them recognition for their efforts from higher commands. These behaviors are admired and encouraged within the Corps.

But what happens when times are hard and leaders are under extreme duress? Every war produces stories of a Marine falling on a grenade to save his squad or of Marines risking their lives to complete their mission while putting the fewest of their fellow Marines in additional danger. That is fairly common fare, but combat and death are not the only types of duress that Marines face. Among the worst of these is long-term captivity.

When Marines are captured by terrorists or become prisoners of war, their responsibilities remain: accomplish the mission, and take care of those around you. The mission may have changed dramatically, and details may be modified, but the essence of leadership does not change. In most cases, the mission becomes to resist whenever possible and to the best of your ability, to never stop looking for a chance to escape, to observe everything for later use, and to give away as little as possible that will be of use to your captors.[39] There is no change, however, in the charge to take care of others in captivity. What is interesting is that by doing so, by acting as an unselfish leader, it is actually easier to accomplish the mission, and it increases the chances of survival of both the leaders and the Marines in their charge.

From January to April 1942, American and Filipino forces defended the Bataan Peninsula from the Japanese. On April 9, allied forces received the order to surrender, only to face the infamous Bataan Death March. An estimated 10,650 people died on this trek to the infamous Camp O'Donnell, many of them murdered when they could no longer stay on their feet or keep up with the others. Of the Death March victims, 650 were American.[40] The deaths continued at Camp O'Donnell. During the first 40 days of that camp's existence, more than 1,500 Americans died. All of these deaths were the direct result of malnutrition on Bataan, disease, and the atrocities committed by the Japanese on the March.[41]

In his memoir, *Never Say Die*, First Lieutenant Jack Hawkins detailed how those captured by the Japanese became demoralized during their captivity: "There were many indeed who became so demoralized that they abandoned every tenet of personal integrity, honor, loyalty, and the accepted standards of human behavior." For many, the discipline began to collapse right from the moment of capture. While some men refused to continue to obey their officers, more disturbing was that many officers surrendered any feeling of responsibility for their men. Hawkins noted that "Military organizations fell apart, and were further broken up by the Japanese in a well-calculated effort to destroy group cohesion and convert the prisoners into an easily dominated, amorphous mass."[42]

Out of these horrors, one group of men stood out from the rest. Of those that perished on the March, relatively few were Marines. According to Gregory J. W. Urwin, Ph.D., of Temple University:

> *"Of the American POWs on the Philippines, U.S. Army POWs experienced a death rate of 42.6 percent for the entire war, while the marines had a death rate of 31.8 percent. For the Pacific theater as a whole, the marine POW death rate was half that of the army's: 22.8 percent vs. 40.4 percent."*[43]

Some of these differences had to do with where the Marines had been before becoming POWs, with somewhat better food and climate. However, even when soldiers and Marines were held captive in identical situations, Marines did better. Lieutenant Hawkins continued:

> *"There was a way to inculcate in men the discipline, loyalty, spirit, mental stamina, and moral fortitude that were called for in the Japanese prison camps. It was the Marine Corps way. I was proud indeed to see that there was no collapse of discipline and group spirit among the*

Marine prisoners. Standards of conduct among the marines were generally excellent, far superior to the norm."

For those Marines enduring the Bataan Death March, leaders—particularly NCOs with more experience—assisted other Marines who needed a hand. They warned against drinking the water, and even purified it with iodine they had hidden in their clothing before surrendering. Once they were in the camps, enlisted Marines joined up with each other as buddies. If one became ill or was punished by having food taken away, their buddies would husband their portions or steal more and give it to their buddy on the sly.[44] Since the Marines were kept near starvation, the unselfish sharing of food was rather remarkable. The punishment for being caught sharing food could also mean losing your own. They did it anyway.

"When your buddy's down, you picked him up, and when you were down, he picked you up," said Corporal Robert Brown. "It happened all the time."[45]

Corporal Onnie Clem, who had been stationed at the U.S. Embassy in what was then Peking, China, volunteered for duty in the Philippines when war broke out there. He was a Japanese prisoner for two and a half years before being rescued by an American submarine.[46] "The Marine Corps had a lot of discipline," he said. "We followed orders and instructions from our officers. There wasn't anybody who fussed with what the officers told them to do."

An army pilot, Lieutenant Samuel C. Grashio, agreed with that assessment. "As a group, the Marines stood up better than most others under the burdens, humiliations, deprivations, and temptations of camp life."[47] Life was hard under the Japanese. The lack of food, medication, and sanitation along with weather conditions and brutality meant a 37.3 percent casualty rate among the thousands of POWs throughout all Japanese camps. By comparison, fewer than 3 percent of POWs held by the Germans during the Second World War died.[48]

"Why should the Marines be different?" asked Lieutenant Hawkins:

> *"They were the same kind of people. I could only conclude that the roots of the difference were embedded in the Spartan ruggedness of Marine training and the fanatical emphasis upon discipline, loyalty, pride, and esprit de corps, which commences for every Marine at the recruit depot."*

Corporal Martin Boyle who was captured on Guam observed:

> *"When a bastard hits bottom he doesn't turn into a nice guy or vice versa. I think it's all there to begin with. This is especially true of a U.S. Marine, where esprit de corps is hammered into his thick skull and ass from his first soul-shattering meeting with a drill instructor."*[49]

Corporal James R. Brown believed that Marine training filled his fellow POWs captured on Wake Island with "the kind of morale that brought most of them back with their morals."

On March 16, 1985, Terry Anderson was kidnapped. He had been a sergeant of Marines during the Vietnam War, and was working as the Associated Press's chief Mideast correspondent in Beirut, Lebanon. On his way home from a friendly game of tennis, he was taken from his car at gunpoint, put into the trunk of another car, and secreted to an unknown location. He had been kidnapped by members of Shi'ite Hezbollah, who hoped they could use him to barter for terrorists who were being held prisoner in Kuwait. Other than about fifteen minutes a day to use the bathroom and to try to wash, he spent his days on a cot, chained to a radiator—for almost seven years.

Anderson recalled his time in captivity: "Marine Corps training doesn't make you invincible, nor does it mean you won't make mistakes." He knew he had made a mistake that morning. Beirut at that time was a dangerous city and no American could afford to let his guard down. Anderson paid dearly for his mistake, as did each of the other hostages. Through his experiences in Lebanon, he knew immediately that his captivity would last for a very long time:

Terry Anderson, a former Marine sergeant and chief Middle East correspondent of the Associated Press, was kidnapped by members of Shi'ite Hezbollah in March 1985. (Photo courtesy of Terry Anderson.)

"So even though I had all that training and all that experience, been through wars, three years in Beirut, which was a really vicious, vicious dangerous place, I made a bad decision. But then, you know what? That Marine Corps training came through. Everybody did what they had to, and we all had to live in the

way we thought we could. But for me, that self-discipline, that firm grounding in reality—you deal with what you have, not in what you want, or what you hope—you deal with what you have. That's what made the difference."[50]

Anderson's captors were very aggressive, violent, and threatening. He was tortured, and that hardened his heart. *Okay; dump on me—I'm just a piece of wood,* Anderson remembered thinking. *You want to shoot me? Go ahead. You're still not getting anything from me. I don't like it here anyway.* Anderson kept his mission in mind: Stay alive. Observe. And resist as you can. When his guards would put the barrel of their guns at his neck and threaten to kill him, Anderson did his best to remain calm. He knew that if they wanted him dead, they would kill him. Since they held the power, that choice belonged to them. But Anderson also had a choice: whether or not to resist. That's what his Marine Corps training taught him. For Anderson, that training was powerful and important. He had to resist with what little ability he still possessed—for himself, and for the other hostages he knew were held throughout the city.

After his capture, Anderson was held somewhere in the rabbit warren of West Beirut. By 1985, he was no longer alone. Five hostages were sequestered in a basement divided into cells by cheap plasterboard. With him were Father Lawrence Martin Jenco of Catholic Relief Services; William Buckley, Beirut station chief of the CIA; Reverend Benjamin Weir, a Presbyterian missionary; and David Jacobsen, director of the American University Hospital in Beirut.[51]

These men also had to deal with repeated death threats. Buckley became feverish, then delusional. He disappeared.

Food was scant. One night, they were fed a mixture of raw meat and cracked wheat called *kibbeh naiya* that made them sick:[52] Gastroenteritis is no joke when you are not allowed to use the bathroom on your own schedule.

Anderson believes his Marine training helped save his life: "They took a seventeen-year-old kid without any discipline or any idea about the world, and gave me a framework. It's the discipline, the self-discipline, the way of looking at things." That training taught Anderson to remain focused in difficult times and impossible situations. He had learned to calm himself and concentrate on the things he *did* have the power to do. He was able to look at each individual terror as a problem to be solved, and create a method to counteract the worst of it. In this way, he could avoid the panic that had demoralized other hostages, and made them much more vulnerable to the guards' capriciousness.

Anderson started a campaign. He began by seeing if he had any influence over his captors, and if he had the ability to change their attitudes, even if only by a small amount. "First thing was to convince these guys that we were hostages—not criminals. We hadn't done anything to them. We weren't here to be punished." Since their goal was to be traded for terrorist prisoners, the health and welfare of the hostages might make a difference in the freedom of one of their friends.

The guards were young and uneducated, and relatively unsupervised. Because their superiors often did not appear for weeks at a time, they could act in any way they chose without fear of repercussion. Anderson worked on them whenever he saw an opportunity: *This is not right. You don't have the right to do this. You have the power but you don't have the right.* He noticed that by repeating this lesson, his guards eased up on their harsh treatment to some degree. Knowing that he could have some influence, Anderson tried to push for better treatment for all the hostages. Over time, his influence got them some books, the chance to listen to the radio, and better food on occasion:

> *"We get a new guard. And he has this bad attitude. I'm sitting in the cell, and he comes in and gives me a kick and says, 'Get up,' with his gun out. I knew that if I said or did the wrong thing, this guy could shoot me in a heartbeat. I turned to Mahmood, a guard who's been there since the beginning, and say 'Mahmood, get rid of this guy or he's going to have to kill me. I've been here three years; I don't need this shit.' And the new guy disappeared! He disappeared that day and never came back. And I thought: 'That was kind of cool.'"*

Since Anderson found that he had the ability to influence the guards, he realized he had an obligation to do so, not just for himself, but also for the other hostages who did not have that kind of power, either because they had not been there as long, or because they did not have the mental attitude to bring risk onto himself for the sake of the others.

> *"It's not a tough-guy attitude, it's a reality attitude. I know they don't want to kill me. I know they will; they're not nice people. But if I play it right, there may be some impact on circumstances. And if I can do this, I need to for myself as well as for these other hostages. And I think that's very much what the Marine Corps teaches. You can*

joke about it, all gung-ho and everything, but it's a way of simply accepting reality and figuring out what to do with it. I know how much it sucks. There it is. But ignoring reality doesn't make it go away."[53]

Anderson had joined the Marine Corps when he was seventeen years old, just three days out of high school. He could have gone to college. He was a National Merit Scholar and he had scholarships available from a number of schools. But money was tight and he didn't feel he had the maturity yet for college. He thought he could prepare for college and gain that maturity through the Marine Corps. After he returned from the recruiting office, his father was very upset with Anderson's decision. But Anderson was confident he had made the right choice, and felt it was the honorable thing to do. Since he had not yet turned eighteen, the Marine Corps held him in California for two months before sending him overseas.

Anderson matured in a hurry. "I got the patented Marine Corps leadership and they do it really well. They take young men, give them responsibility, and teach them how to be leaders." This paid dividends for him in the Corps and later after discharge. When he joined the Associate Press after graduating from college, the AP discovered that he could operate effectively in dangerous situations. Not all journalists had the ability to function when bullets were flying. Anderson got the stories they wanted without getting killed and without endangering other journalists. "If you get killed," Anderson said, "you can't file your stories."

Anderson's early assignments in the Marine Corps included a stint in Japan where he was supposed to remain for three years. He made rank very quickly and reenlisted while in Japan serving with the Far East Network, which was a Joint Service organization run by the Air Force, with about six Marines. "We were nominally assigned to the Marine Barracks. I get this call from the first sergeant and he says, 'You got orders.'" So in 1969, Sergeant Anderson came back to the States, attended the Combat Correspondent's Orientation Course, and then was shipped off to Vietnam as a Combat Correspondent.

His first boss was a retread from World War II. His second boss believed in getting as deep into the bush as you possibly could. And that's what they did. It was an odd existence with contradictory experiences. The combat correspondents lived in the Danang press center, where there were restaurants, showers, and air conditioners. There were even civilian correspondents in the front of the center with the officers. He would get assignments to go into the field from his master gunnery sergeant, grab his pack, go out to the hospital and then hitch a ride on a medevac helicopter out to where the stories were. The reason the medevac was going in was because people had gotten shot—he was leaving the comforts of Danang for the heart of combat.

Even though he was a journalist, like all Marines, he was a rifleman first. When joining a unit in combat, even though he wasn't a regular part of that unit, he was put to work where it was most needed. "They'd stick you in the shortest squad, hand you some mortar rounds—somebody's got to carry the mortar rounds, and you're the new guy— they don't care that you're a sergeant," Anderson remembers. Eventually, he would return to Danang and all its comforts to write and file stories about what he had experienced and observed. Then the cycle would repeat again. Anderson recalls: "We spent more time in the bush than anybody."

After the Tet Offensive in 1968, the strategy for Marines in Vietnam changed. Anderson called it trolling: "Send out the Marines, a platoon or two at a time, walk them through the hills and wait until somebody shot at them. Then they could bring in the artillery or whatever.

"And you get to where keeping your buddies alive— that's the whole game."

Eventually, Anderson said goodbye to the Marine Corps and enrolled at Iowa State University, studied broadcast journalism, and graduated in 1974. He then joined the Associated Press. He traveled to hot spots around the globe, always heading for action. He felt—and still feels— that good journalists want to go where the stories are, no matter what the dangers. The stories that come out of war zones are important and necessary, both to educate and inform the public and to get the story on the front page.

The war zones are dangerous, however, and not everyone has the advantage of Marine Corps training and experience. When Anderson held the senior post in Beirut, he knew he had to take care of his journalists and make certain they had the skills needed to function in dangerous situations. While he was in charge, his people filed exceptional stories, but not one journalist was killed. Anderson credits his experiences as a Marine and in Vietnam as crucial to his success in both areas: "The Marine Corps, Vietnam, having actually been in a real war, that was all important, and helped make me a better journalist."

Anderson can't pin down exactly what it is that he learned from the Marine Corps that helped him survive as a hostage. It wasn't about specific training for specific instances; it was more about making him the kind of person who could get through it, and help others survive. Anderson said:

> "What's cool is that whether you're all into the Corps or not, the system still works, and it works in extreme circumstances, which is what it was designed to do. They've been doing this a long time and been doing it very well. Psychologically, they're very sophisticated. They know exactly what they're doing. And it's far beyond the brainless 'teaching people not to think.' You can't do that. The average Marine may be handling a million, two million dollars worth of gear—and making snap decisions. You've got to be smart. The average Marine has a great ability to take initiative and think independently."

Anderson's ability to think and take that initiative allowed him to unselfishly mold the environment for all the hostages.

On December 4, 1991, Anderson was released. He works now to help other journalists who don't have the advantage of Marine Corps training, continuing his unselfish labors to protect others. Working with the Committee to Protect Journalists (CPJ), a major organization that strives on behalf of journalists in danger, he fights to protect writers without military training who have no idea of how to deal with dangerous situations. He also helps to teach them how to write about war.

"If I send you to Wall Street to cover a story, I have a reasonable expectation that you're going to know what a put and a call is, and the history of the stock and the bond. That's basic professionalism. But we've got people coming in to cover war stories who don't know the difference between an M-16 and an M-14, let alone all the duties of a sergeant or a major, who don't know the functions of a platoon or a company, and have next to no idea of how a military or paramilitary organization operates. That doesn't make for a good story, and it's dangerous, for the military as well as the journalist."[54]

In every war, and in many undeclared conflicts, Marines have needed to survive captivity. During the Korean War, about 7,200 Americans became POWs; 221 of them were Marines. Through a combination of discipline, humor, and unselfishness, only 12 percent of captive Marines died in enemy hands. The death rate for Americans of other services approached 39 percent. "Without USMC training I would never have lived," said Captain Charles Harrison, who had been a private first class on Wake Island when he was taken prisoner—the first time.[55] He was captured again during the Korean War, managed to escape, and continued on to serve two tours in Vietnam.[56] Said Harrison, "I think I know better than most how sweet free air tastes. I'm not yelling uncle, but two times as a P.O.W. are too much."[57]

Although early reports were published stating that no American POWs attempted to escape from captivity during the Korean War, Marine Corps records show that in addition to Harrison, eighteen Marines and a U.S. Army interpreter attached to the 1st Marine Division escaped in May 1951 after several months as prisoners.[58]

The Marine POW record in Korea was superior. When Congress investigated the issue of prisoner-of-war conduct after the armistice, the Senate report summarized:

> *"The United States Marine Corps, the Turkish Troops, and the Colombians as groups, did not succumb to the pressures exerted upon them by the Communists and did not co-operate or collaborate with the enemy. For this they deserve greatest admiration and credit."*[59]

Unselfishness taught during Marine training allowed these men to survive long-term captivity and to help fellow prisoners do the same.

By sticking together and unselfishly helping each other, Marine Corps-trained long-term captives came home with their lives, their dignity, and an enduring sense of brotherhood. Corporal Robert Brown spoke for more than just the Wake Marines when he observed: "I don't know of any other unit cohesion that works as well as this did. In the very bitterest days we had, this group got through by ... one of the greatest cases of friendship the world has ever known."[60]

Unselfishness is an important trait of Marines, a trait that has helped many survive unthinkable situations. Bearing and initiative, along with selflessness, go to the heart of Marine Corps training. So does dependability.

Chapter 4

Dependability

They wish to hell they were someplace else,
and they wish to hell they would get relief.
They wish to hell the mud was dry,
and they wish to hell their coffee was hot.
They want to go home.
But they stay in their wet holes and fight,
and then they climb out and craw through minefields
and fight some more.
—Bill Mauldin (1921–2003)

A midst the stress and chaos of combat, there is often no telling how people will react. A hero one day may be a catatonic wreck the next. Some would say that's perfectly understandable. Marines say that's totally unacceptable. Marines demand dependability in all situations—on or off the battlefield. Many former Marines now in the business world would testify that the stress they face may not be as bloody but that it is just as brutal. And they would say that success in the boardroom or on the factory floor hinges on the same sort of dependability that was demanded from them in uniform.

It is easiest to understand the importance of dependability among leaders by looking at examples in battle where the failure to step up—the failure to do what's required—can lead to bloody disaster. There are examples throughout

Marine Corps history, but truly shining examples can be found among leaders in the Marine Brigade sent to France after America entered World War I in 1917 and sent combat troops to bolster allied forces struggling with a stalemate in the trenches of the Western front. When the Marine Brigade entered combat for the first time in World War I, some old leaders fell by the wayside and many new ones emerged in arduous fighting like the attack on Blanc Mont in fall 1918. Fights like Blanc Mont reiterated the importance of dependability in the ranks where the battle was brutal and fought at very close range.

Blanc Mont in France is named for the chalk-white limestone soil of the area. German forces had constructed dense defensive positions in the area and held firmly against repeated allied assaults beginning in 1914. The decisive battle at Blanc Mont began four years later in October 1918, when the American Expeditionary Force's 2nd Infantry Division was given the difficult task of helping to do what French forces had been unable to accomplish for four long years: drive the Germans off the battlefield and hold the captured ground against inevitable counterattacks. German forces were dug in on high ground overlooking long, open slopes that were fertile killing grounds for defending machine gunners and artillery forward observers. Captain Dave Bellamy, an officer in the Marine Brigade attached to the Army's 2nd Division for the battle recorded a wry comment about the mission at Blanc Mont in his diary: "The French have certainly given us a hard nut to crack."[61] It was an understatement that quickly became obvious to Marine NCOs like Sergeant William March Campbell who were gearing up mentally and physically to cross the killing fields and attack Blanc Mont Ridge as allied artillery and aircraft pounded the area.

Campbell was in many ways typical of the men who came from civilian pursuits, flocked to the colors and went to war in France with the Marine Brigade in 1917. There were tough, hard-edged men with little education who came from farms and factories, and there were those like Campbell who left promising scholastic or professional careers to get in on the action. Campbell was born on September 18, 1893, the eldest son of a poor family in Mobile, Alabama. One of eleven children, March had a Spartan upbringing. By the time he was fourteen, the family moved to Lockhart, Alabama. There was no high school in the area, so he found work in the office of a lumber mill.[62] Two years later, he'd saved enough money to move back to Mobile where he found work in a local law office. He developed an affinity for legal work and saved enough money at his job to eventually enroll in a course which gave him the equivalent of a high school diploma and allowed him to enroll in law school at the University of Alabama. Money ran out before he could finish his degree, so he moved to Brooklyn, New York, and found work as a clerk in the law firm of Nevins, Brett and Kellog.[63] He was twenty-four years old and settling in to what appeared to be a solid career when America entered World War I and began to send troops to France. Campbell decided he needed to be there and do his bit for the cause. He selected the Marine Corps as the outfit most likely to get him into the action. He found a recruiter and enlisted on July 25, 1917.

William March Campbell
Official Marine Photograph,
taken 1918. American
Marine and author, he was
highly decorated for his
actions during World War I.
(Public domain.)

After basic training, Campbell was sent to France where he arrived in February 1918. He was assigned to the 43rd Company, 2nd Battalion, 5th Marines as a replacement. It was a prestigious assignment for a man looking to get into the fight. Campbell's new outfit was one of the most celebrated fighting units in the Marine Corps. By the time Campbell joined 2/5 the battalion had already acquired a motto derived from the response one of its officers gave to a French ally who urged them to retreat as they were headed for duty in a collapsing battle front. Legend has it that Captain Lloyd W. Williams laughed at the French officer's warning and said: "Retreat? Hell, we just got here!"[64] From that point and into history, 2/5 immortalized the words "Retreat Hell" as a revered motto.

Campbell and fellow replacements who joined 2/5 in France had a fighting image to support by their efforts and they did their best in bloody battles at infamous places like Belleau Wood and Soissons before they found themselves facing Blanc Mont. Along the way, Campbell earned promotion to sergeant but had done mostly clerical work within the battalion. That was about to end at Blanc Mont.

The attack went forward with troops from the Marine Brigade—including Campbell and the rest of 2/5—surging uphill closely following a rolling artillery barrage. They fought their way through lines of German riflemen and machine gun positions under incessant pounding from German artillery and mortars. It was bloody, brutal slogging as Marines cleared hard-points, wiped out artillery forward observers and drove German defenders out of reinforced trenches. Eventually they gained the crest of Mont Blanc and began to consolidate their positions for German counterattacks that were sure to come.[65] The Marines looked around for the French forces committed to the attack who were supposed guard the Marine flank. And that is when times got really tough for Campbell and the surviving Marines at Mont Blanc.

The French units never left their trenches for the attack and the Marine flank was wide open. It was as obvious to the Marines atop Mont Blanc as it was to the German forces already gearing up to re-take the high ground. Another bloody fight developed as the Marines tried to hold the ground they'd taken at such high cost and the Germans tried to lever them out of their former bastion.

At that point in the battle, Campbell—by now an NCO with a reputation for dependability and for doing what needed to be done with or without orders—was lending a hand with the wounded in a slap-dash medical clearing station on the hill. The Germans struck hard in a direction that would take them right through the aid station. Campbell could see that, and he could see what needed to be done to prevent it. An excerpt from his Navy Cross citation describes what happened next:

"...while detailed on statistical work, Sergeant Campbell voluntarily assisted in giving first aid to the wounded.

*On 5 October when the enemy advanced within 300
yards of the dressing station, he took up a position in the
lines, helping in defense. Although twice wounded, he
remained in action under heavy fire until the enemy had
been repulsed."*

What remains unsaid in the typically terse citation is
that Campbell was a virtual demon in defense of the
wounded and the position at Blanc Mont against swarms of
German attackers. He knew it was do or die for him, the
wounded men, and the corpsmen treating them. He also
knew their lives depended on someone taking swift, decisive action. There were no orders issued at the time. There
was only Campbell and the others still capable of fighting
to prevent disaster.

Campbell was dependable. He was where he needed to
be and doing what needed to be done. His efforts contributed significantly to holding the line and winning the fight
at Blanc Mont. That fight was a key to driving the German
Army from the Champagne Region of France and led
inevitably to victory in World War I.

Campbell developed a significant reputation for reliability in his unit even prior to Blanc Mont. In other battles he
was a dependable hand to the extent that he eventually
wound up adding an Army Distinguished Service Cross
and the French *Croix de Guerre* to his military resume. In
tense situations from one flank of the western front to the
other, Campbell saw work that needed to be done, regardless of his official assignment, and he got it done, reliably
every time and all the time. At Blanc Mont, he wasn't a
medical corpsman, but he saw Marines in need of medical
care and he pitched in to repair battle damage. He was

assigned regularly to do "statistical work," which matched his education and experience in civilian life, but he didn't let that keep him from picking up a weapon and making a big contribution on the fighting line. He was dependable in a crisis. He saw what needed doing and he did it.

Campbell was clearly the type of Marine NCO who demonstrated dependability in dire circumstances, but his brand of reliability is important in leaders on and off battlefields. Leaders deal every day with more mundane challenges. But that doesn't detract a bit from the importance of dependability to their mission in the eyes of both superiors and subordinates. In large measure the success of any organization hinges on the dependability of its leaders. It is often difficult to gauge with a yardstick, but there are some marks that assist in measuring that sort of leadership. A reliable leader develops a capable team through training that fosters utter reliability and by personal example proves he can be relied on by superiors to accomplish the mission and by subordinates to take care of them along the way.

Dependability obviously incorporates technical expertise. Teams usually can rely on an outfit that knows how to get a job done but there is much more to it than that. What happens if the situation demands excellence or exceptional performance from people who don't have that particular expertise? As it happened with Sergeant Campbell at Blanc Mont, what happens when a leader with some sort of non-combat expertise is suddenly thrust into the cauldron and required to perform as a line infantrymen? What happens when a helicopter aircrew is forced down or shot down by enemy fire and has to take their place in the fighting line? Are they an asset or a liability? Can they be relied on to operate outside their technical expertise and the pages of a manual? Sometimes problems arise that are within the ability of a team to solve, even though there isn't a policy or procedure in place to handle it—if they have been empowered to act in those kinds of situations. <u>Too many small unit leaders in business are not given the authority needed to act decisively when presented with a problem, and so they pass the problem on to others.</u> Because of this inability to problem-solve, they can get the reputation of not being dependable.

However, if small unit leaders are given that trust, then they can handle a variety of problems right on the spot. And that's the goal for Marine leaders, especially Marine NCOs who often face such situations. Senior Marine leaders know that, and they rely on their NCOs to act effectively and perform admirably outside occupational parameters in difficult situations.

One of the keys to Marine Corps leadership is that inherent trust in NCOs, small unit leaders who are given the authority to solve problems at lower levels in the chain of command. In fact, "it's at the lowest level of Marine leadership—the corporal—that the Corps has focused the lion's share of its skills development efforts in recent years."[66] That doctrine is what marks Marine NCOs as people who can be depended on to solve problems on the ground whenever and wherever they occur.

That trust and training pays remarkable dividends for Marines who leave the active ranks to fight different kinds of battles in civilian pursuits. Fortune 500 companies in

American industry are shot through with Marines who have brought their training and leadership perspective to the world of business. And it is often a brutal battle for survival that calls on their background and training to get difficult jobs done among civilians who just don't quite understand—much less agree with—the Marine Corps concept of dependability.

Take for example the case of Bradley J. Hartsell who spent ten years in the Marine Corps before he left the uniformed ranks to fight new battles in a worldwide technology firm. Hartsell served on active duty from 1978 through 1988, held a top secret security clearance, and left active duty as a staff sergeant. While in the Corps he served in several billets including platoon sergeant and water safety survival instructor, but his official job description was the high-tech electronic cryptographic repair technician. It wasn't a line infantry fighting assignment, but Marines like Hartsell had to be utterly dependable in what they did.

There was no room for delays or whining about being in a rear echelon pursuit. The equipment they maintained and repaired was vital to Marine Corps communications. They had responsibility for truly sophisticated equipment that kept communications working and beyond the ability of others to intercept or decode. It was sophisticated gear that demanded bright, talented people to understand and maintain. It also involved a leadership challenge for NCOs like Hartsell who constantly had to remind their Marines that they may be back behind the fighting line but they had a vital job to perform in the overall scheme of Marine Corps operations. The key to success in Hartsell's Marine Corps job was total dependability.

"If you say you are going to do something or be somewhere then, by God, do it!" Hartsell is still adamant about
that aspect of leadership:

THIS IS HUGE

*"If you tell someone I'll get it done in two weeks but you
deliver in one week, they will be very appreciative. But if
you tell someone I'll do it in one week and you deliver in
two weeks, then you will be branded as someone who
cannot be depended on to do what they say they will do."*[67]

Hartsell carried this principle into civilian pursuits from
his time in the Corps. He believes in underpromising and
overdelivering, and he firmly believes that good leaders
reserve their promises for those times when they absolutely
know they will be able to deliver. He learned in his experience as a Marine NCO that if he was unsure, then he
would not promise what he might not be able to deliver.
This philosophy has earned him a good career in a fast-
paced and highly competitive civilian job. He didn't get
there by simply spending time or milling around at climb-
ing a corporate ladder.

Former Sergeant Hartsell, like Sergeant Campbell before him, continued his education and constantly studies
his business so that he will be the go-to guy, the person
who can be depended on to get any job done in a pinch.
While stationed on Okinawa, Hartsell began to study the
Japanese language and, in the process, developed a serious
avocation for Japanese history, especially the history of
Imperial Japanese Forces during World War II. In pursuit
of esoteric knowledge, he excavated Japanese tunnels
systems and retrieved rare artifacts and mummified human
remains on Pacific islands that were battlegrounds during

World War II. It might seem like a sidestep, but Hartsell considered it invaluable in explaining Marine Corps tactics and motivation to his Marines. He saw his studies as part and parcel of making him a well-rounded individual who could be relied on to handle unique situations.

Out of uniform, Hartsell continued his formal education at the University of Michigan where he received a Bachelor of Arts degree in Japanese Studies. Following graduation, he built a successful business career and attained a Masters Degree in business administration. That education and his Marine Corps ethic led to a high-level position with a firm doing business throughout the Far East, and he has lived and worked in Shanghai, China, for almost a decade. Every day, Hartsell says, he's relied on his Marine Corps training and a hard-won reputation for dependability to survive and succeed in the demanding fast-paced pursuits of international commerce.

Hartsell says that dependability ties into the "3-11 Rule." The 3-11 Rule is a common business gauge that maintains on average, a happy customer will tell share his satisfaction with only three people. An unhappy customer, on the other hand, will tell eleven others that he's not satisfied with a product or service. That's an important concern for people struggling for supremacy in the competitive arena of international commerce. If a company can't be depended on to keep its customers happy, the bad news circulates rapidly, and there's a damaging ripple effect. Once a company has been tagged with a reputation for a lack of dependability, getting back in the good graces of anxious customers is a tough, uphill fight. It's a lesson Hartsell learned from his earliest days in the Marine Corps:

"If the senior NCO—the squad leader or platoon ser-
geant—tells Private Jones he'll get his payroll squared
away before the next pay period and then he doesn't do
that, regardless of the reason, imagine how Jones feels.
Imagine how the other Marines in his squad or platoon
feel when they hear that Jones has been let down on a
promise. It's devastating to morale and it casts major
doubt in the minds of the other men about the dependa-
bility of the leadership of their unit."[68]

There's an instructive intersection between Sergeant
Campbell's experiences in World War I and Sergeant
Hartsell's experiences in industry that focuses the importance of dependability. Consider the example of developing a reliable general purpose machine gun for the war in
Vietnam. If anything needs to reliable in combat, it is the
weapons used by the soldiers involved in the fighting, and
in modern warfare, that centers on the squad machine gun.

In the early 1960s, American forces were facing both a
looming Soviet threat in western Europe and burgeoning
communist insurgencies in Southeast Asia. It had become
clear to the people tasked with doing whatever fighting
might develop in either hot-spot that the U.S. .30 caliber,
infantry-support machine gun, which demanded a tripod
for effective fire, was not going to be sufficient on modern
mobile battlefields. The call went out to American industry: develop a machine gun for infantry combat that was
totally reliable and could be used at the squad-level by
troops on the fly. It seemed like a simple mission but the
resulting responses were somewhat less than reliable. After
a tumultuous period of trial and error, the American

military decided the M-60 machine gun came as close as anything industry had to offer in filling a pressing need among American forces and their allies.

It did not take armament industry leaders long to discover that what had been good enough during World War II was not going to work on modern battlefields. Times were changing and industry had to change with them. World war was a genuine possibility and limited war on exotic battlefields was a reality as industry "began to discover that a lot of the things they had done had been done very badly indeed; that a lot of things that they had always maintained couldn't be done could be done, and were better than the old ways."[69] This revelation applied to an unlimited number of military contracts but a focus on the general purpose machine gun is instructive. The major concern was producing something that was utterly reliable and providing that to the American military demanded dependable producers who understand both the requirements and the importance of the problem. The first crack off the industrial bat was the M-60 machinegun which had been issued—but not necessarily proven—to troops heading for increasingly violent combat in Vietnam. It was a weapon designed by committee and it had serious teething problems among the troops who had to rely on it in bloody encounters with insurgent Viet Cong or conventional North Vietnamese forces.

U.S. Army and Marine infantrymen were having serious reliability issues with the M-60. Its maintenance requirements were well beyond expectations in field conditions, and it was prone to jamming, especially when it was dirty from excessive firing. It would malfunction when sand or

dust got into the mechanism and it was a constant maintenance nightmare. "The barrel change lever could not be
worked without asbestos gloves."[70] On test ranges, the M-
60 could fire thousands of rounds without jamming but
those were ideal conditions and conditions in the jungle
were far from ideal. A Marine sergeant who served in
Vietnam described how he and his unit handled a problem
with feeding heavy ammunition belts through the M-60 in
assault fire situations. The gun fed smoothly when it was
stable and fired from a bipod or tripod, but it tended to jam
when gunners tried to use it in assault fire. The Marines
rigged a fix using a C-ration wired to the feed-tray to
reduce the angle of a belt of heavy rounds feeding into the
gun and made it work. Meanwhile, complaints were
communicated to American industry and armament
engineers went back to their drawing boards.

With reliability and dependability as guideline, American engineers eventually came up with the M-60E1 version
of the weapon that solved the problems being encountered
by Marines and soldiers in the field. It wasn't an easy deal
for industry engineers who were facing corporate and
bureaucratic obstacles in solving pressing problems that
were costing American lives on far-flung battlefield.
Solutions demanded engineers and industry leaders who
were both technically proficient and—above all—
dependable. This wasn't simply a matter of profit and loss.
American and allied lives were at stake.

In the case of the M-60 machinegun, American industry
stepped up to bat and eventually hit a homerun—unlike the
situation with the M-16 rifle which failed miserably in
combat and never worked reliably until well after the war in

Vietnam where it was first fielded—but it was a classic case of failure of dependability throughout American industry. Up and down the design and production pipeline, people failed to be reliable and that cost lives. Men like Sergeant Campbell and Sergeant Hartsell would have understood the problem and would never have tolerated the situation.

Situations like the one that developed with design or the M-60 machinegun and the M-16 rifle have not gone unnoticed by industry. In 2007, the Corporate Executive Board's Learning and Development Roundtable—a group of executives mostly from large firms who are responsible for cultivating leadership talent—discovered a paradox among newly-appointed leaders. They found that these leaders focused on the quick-wins, trying to score points with their bosses by quickly acquiring successes.[71] The problem with this strategy is that by trying to gain a variety of achievements quickly, they overextended their commitments. Trying to have a long list of accomplishments, they end up sacrificing the quality of what they do achieve. To be dependable, attention must be paid when a commitment is given. The teething problems with the M-60 machine gun and the M-16 rifle are classic examples of what happens when industry takes a less-than-dependable attitude to military requirements and there are hundreds more to make the point. Dependability is crucial, and producers who understand that succeed. Those who are not dependable usually fail miserably, often costing lives and fortunes to those who depend on them.

Leaders—military or civilian—must work within an organization and that means surviving and succeeding within a system. No system is 100 percent reliable, but dependable leaders can make that system much closer to foolproof. That's the lesson Marines like Sergeant Campbell, Sergeant Hartsell, and the engineers who eventually solved the problems with the M-60 machinegun learned and internalized. Dependable people can keep a system functioning and moving forward because they are there where they are needed. Dependability is not inherent in systems. It is the human quality and contribution that makes systems work against all odds.

This is the lesson that these people can teach us about dependability. The M-60 served admirably and reliably with American military forces for years until it was eventually replaced by a better weapon. Hartsell continues to lead and innovate in industry where he is considered among the most reliable and dependable executives in a highly competitive enterprise. And Sergeant William Campbell left a legacy of dependability that lives on in popular culture.

Following the Armistice of World War I on November 11, 1918, Campbell stayed in Europe with the infamous "Watch on the Rhine," troops held in position along the Rhine River as a blocking force in case the defeated German military in 1918 decided to give it another try. The assignment gave victorious allied troops something a mission while they waited for transportation home and ultimate demobilization. During the interim, the American Expeditionary Force created a voluntary education program for those still in Europe, and Campbell took advantage of it. He was detached to attend classes at Toulouse University where he spent several months studying journalism.[72]

Shortly after he returned home for discharge, Campbell landed a job with the newly founded Waterman Steamship

Corporation in his hometown of Mobile, Alabama. His dependability was quickly noted, and he quickly rose to positions of leadership within the company. While developing this career, he continued the writing he began in France with a personal diary. Using those notes and wartime letters he had written to an elder sister during the war, he started constructing character monologues which collectively became his first novel, *Company K*, published in 1933 under the pen name William March.

That work was reviewed by *New York Times* veteran war correspondent Arthur Ruhl who called the book a classic piece of World War I fiction and noted in his opinion: "The outstanding virtues of William March's work are those of complete absence of sentimentality and routine romanticism, of a dramatic gift constantly heightened and sharpened by eloquence of understatement."[73]

Encouraged by such praise, Campbell left a post-war executive position in 1937 to devote full time to writing. His last novel, *The Bad Seed*, published in 1954, became his most notable success. The paperback edition sold more than a million copies and was adapted as a motion picture and a Broadway play. Unfortunately, Campbell never learned of these successes. He died two months after the book was published.[74]

Campbell's post-war success clearly indicated that he understood and lived a life based on being dependable. He understood the importance of that leadership trait. He understood—in and out of uniform—that great leaders need to be dependable. He realized, as did Hartsell and hundreds of Marine NCOs who followed his example, that subordinates in and out of the Marine Corps prefer a leader on whom they can depend.

And he knew that dependable leaders do what needs to be done—even if it's not their job at the moment—and that they do it consistently in stressful situations. They develop dependable teams by trusting them to make decisions and solve problems. They have consistency in crisis, and maintain their bearing. They do not over-commit. They know when they can't do something, and then make other arrangements to get it done anyway. And they work at being fully qualified for their job, expanding their knowledge beyond the job. It's a measure of their success that they are dependable in numerous situations.

Chapter 5

Endurance

If a task which normally requires a million men can be carried out by one man, this one man possessed psychologically an all but infinitely higher endurance than any single man out of the million ... one Achilles is worth a hundred hoplites.
— Major-General John Frederic Charles Fuller
(1878-1966)

Marine NCOs have borne the brunt of some horrific battles over the centuries and, like Roman Centurions in earlier bloody conflicts, they have been the stalwarts who anchor the fighting line under pressure. A classic example of the monumental endurance required to stand, fight, and inspire others to hold can be found in the actions of Sergeant "Manila John" Basilone at a shell-scarred ridge on Guadalcanal in 1942—and beyond.

Basilone always seemed to have more energy than common sense, even from the beginning, according to those who knew him in his hometown of Raritan, New Jersey. After finishing at St. Bernard's Parochial School at age fifteen, he went to work rather than high school, which his mother would have preferred. By eighteen, he was restless and joined the Army. A champion boxer, he eventually was shipped off to the Philippines.[75] After three years, he was honorably discharged and returned to Raritan, where he worked for a while in a chemical plant. When he became convinced that the United States would enter World War II, he wanted a piece of the action. So he joined the Marines.

The Marine Corps was a good fit for Basilone, and as he moved through stints at Quantico, Parris Island, and New River (now Camp Lejeune), he became a corporal and then a sergeant, assigned to the 7th Marines.[76]

The 7th Marine Regiment was the first unit of the 1st Marine Division to leave the United States. It was expected to be the first to face the Japanese enemy—probably on Samoa—so some of the best weapons experts in the Marine Corps scrambled for an assignment to the regiment. Machinegun specialist Basilone was one of them, and during this assignment he developed an even loftier reputation as the best among a cadre of veteran Marine machinegunners. "Johnny was smart, full of confident leadership and knew his weapons," said Sergeant William Lansford, who trained under Basilone and served with him later on Iwo Jima. "But the big Browning machine gun, .30-cal., Model 1917A1—that splendid, water-cooled, defensive gun—was his meat."[77] By the time the Japanese attacked Pearl Harbor and plunged America into World War II, Sergeant Basilone and his machinegun section were ready for whatever the Japanese could bring to ground combat.

The regiment left for the Pacific on April 2, 1942, and began training in jungle warfare on Samoa. On September 18, 1942, they landed on Guadalcanal in the Solomon Islands, joining their parent division that landed in August to commence the first allied offensive of the Pacific War.[78] The Solomons were prime real estate for both sides: The

Japanese needed the land for the airstrip they were building, an aerial extension of their major base at Rabaul. The Americans wanted it as a launching pad for further offenses in the Pacific and to prevent further enemy encroachment toward Australia. The side that held the Solomons could also control vital sea traffic in the South Pacific.

The United States Marines were determined to anchor their tenuous foothold on the small airstrip they named "Henderson Field" after a Marine aviator killed in the opening days of the war. The Japanese were equally determined to drive them into the sea. During the protracted battle which lasted six months, the struggle to hold Henderson Field came to a bloody climax on the night of October 24, 1942.[79]

It was dark and quiet on that rainy evening as Basilone checked with the men in his machinegun section. Everyone and everything seemed ready. Although the gunners were soaked and many were suffering from malaria, the men of Basilone's section used their ponchos to keep their guns and ammunition dry.[80] Staying alive that night was a higher priority than staying comfortable. Their M1917A1 heavy machineguns were the key to survival on what would become known as Bloody Ridge.

Although the media had taught civilians back home that the jungles on Guadalcanal were hot and steaming, in reality "many who served there still recall the cold discomfort of their dungarees plastered to their bodies by chilly rain and the icy metal of their weapons as they lay in slimy mud waiting for the enemy to move."[81] A bigger problem was visibility in the inky dark that covered Basilone and his gunners squatting behind their weapons.

Through the dense, dark jungle in front of the fighting line, the Japanese attacked, screaming and throwing hand grenades. Continually regrouping and re-engaging, they seemed unstoppable. Basilone tore into them with his machineguns, piling up bodies in front of his positions. As one Japanese element broke through the line and threatened an adjacent C Company machinegun position, Basilone hefted one of his guns onto his back and took some of his men to relieve the pressure on that flank.[82]

While his men set up the gun and began firing at the enemy, Basilone worked on the weapons that had been abandoned or had jammed. Outside the wire, the Japanese were moving through the rain. Basilone rolled from gun to gun, "shooting up each successive belt as soon as it was fed into the breech and snicked into place."[83]

Private First Class Nash W. Phillips, of Fayetteville, North Carolina, was with Sergeant Basilone on Guadalcanal. "Basilone had a machinegun on the go for three days and nights without sleep, rest or food," he recounted. "He was in a good emplacement, and causing the Japs lots of trouble, not only firing his machine gun but also using his pistol."[84] At the end of the battle, only three Marines from Basilone's machine gun crew were still standing, but they had stalled the best efforts of an entire Japanese regiment.

Basilone may have been still standing, but he endured only with a pair of burned hands. Machinegun barrels heat to very high temperatures under high volumes of fire. To facilitate handling hot guns, Marines were issued asbestos gloves for changing barrels and other necessary manipulations—but Basilone's gloves had been lost in the chaos on Bloody Ridge. To keep his steaming guns up and running, he used his bare hands. At the height of the battle that night, he'd even cradled a smoking hot weapon in his arms and with the heavy tripod still attached, Basilone fired it to kill a wave of swarming Japanese soldiers. The gun seared painful burns on his hands and forearms.[85]

Phillips recalled thirty-eight Japanese bodies piling up in front of the line. This wall of flesh provided cover for the enemy and had to be taken down so gunners and riflemen could fire on their following assailants. Stories differ as to whether Basilone cleared the bodies away or if he gave the order to do so to one of his men. Either way, the wall was collapsed that night by the Marines, who ran through fusillades from both sides to clear the bodies and open fields of fire. Phillips lost a hand fighting next to his Sergeant. He was surprised to see Basilone appear next to him in a field medical station a little while after dawn:

> *"He was barefooted and his eyes were red as fire. His face was dirty black from gunfire and lack of sleep. His shirt sleeves were rolled up to his shoulders. He had a .45 tucked into the waistband of his trousers. He'd just dropped by to see how I was making out; me and the others in the section. I'll never forget him. He'll never be dead in my mind!"[86]*

In the words of the United States Marine Corps doctrinal manual, *Warfighting*, Basilone demonstrated:

> *"a state of mind bent on shattering the enemy morally and physically by paralyzing and confounding him, by avoiding his strength, by quickly and aggressively exploiting his vulnerabilities, and by striking him in a way that will hurt him most."[87]*

That is the official assessment of what Basilone did that night on Bloody Ridge. There is no doubt he also showed exceptional endurance.

John Basilone received the Medal of Honor for his actions at the battle of Guadalcanal. He was the only enlisted Marine in World War II to receive both the Medal of Honor and the Navy Cross, which he was awarded for his actions at the battle of Iwo Jima. (Public domain.)

Physical endurance is the most straightforward of the types of endurance Basilone demonstrated throughout his career, and it is clearly displayed by his actions on Guadalcanal. Throughout the battle, he had the ability to keep doing the same thing over and over for as long as necessary. Marines require extreme physical endurance to accomplish their missions, and they train constantly to keep their bodies in the best shape possible.

A Marine's life may depend on his or her ability to move his or her weight—and maybe the weight of another Marine—around the battlefield.[88] NCOs in particular not only need to train and develop this endurance for themselves, but also for those who look to them for leadership. First they learn to do it, and then they teach it to others.

As Gulf War veteran Sergeant Sean Bunch said in an interview, "It is every Marine's responsibility to remain in exemplary fitness to be as ready for combat as possible."

But what about those who don't need to stop enemy combatants, in the dark and in the rain, with only a few others who are exhausted and malaria-ridden? What sort of endurance do they need? The answer depends on missions and goals.

Marines today train toward specific goals in endurance in much the same way Sergeant Basilone trained his men for the ordeal they faced on Guadalcanal. This training doesn't require fancy equipment or costly gym memberships. The best training is specific to what you'll be called on to accomplish. The Marines test individual combat readiness with the Combat Fitness Test. Most of the movements associated with the Marine Corps' Functional Fitness Program, a program designed to imitate combat functions and to prepare Marines for the CFT, can be performed with a filled sandbag, a rock, a filled five gallon water jug, a tire, or an ammo can filled with sand.[89] For example, fancy running shoes are great, but Marines don't get to wear them in combat. So while they're terrific for having correct form and avoiding injuries during long running training sessions, at some point Marines need to train in their boots. Sergeant Major Randall Kenney, the base sergeant major at Marine Corps Logistics Base in Albany, Georgia, thought about the ideas behind functional fitness and the CFT, and he thinks it is an excellent idea:

*"I might be able to run fast and straight in my running
shoes. But when I'm saddled with a load, I might not be
the go-to person. Then again, I might not be the fastest
one in my go-fasts, but when I get my pack on, I'm going
to get my nose down and get the job done."*[90]

Although Basilone's battle was in October, his family didn't
hear from him again until the following June, after he'd
been presented the nation's highest award for valor in
combat. At that time, a letter came on cheap note paper, in
John's schoolboy hand: "I am very happy for the other day I
received the ... Medal of Honor .... Tell Pop his son is still
tough. Tell Don thanks for the prayer they say in school for
us ...."

On a bright September Sunday in 1943, Basilone got his
welcome home. At the estate of Doris Duke Cromwell,
thirty thousand fans assembled to greet him: mayors,
judges, former Governors and Senators, and a movie star
with an upswept hairdo, who kissed Basilone on the
mouth. His picture hung in shop windows, alongside
General MacArthur's. His portrait was hung in the Raritan, New Jersey town hall.[91]

Basilone was sent on a Treasury-conducted war bond
tour. Marine officers who accompanied him found that he
was still steady, modest about his honors, and anxious to
get back to his outfit.

Basilone wasn't cut out for the high-life of a conquering
hero. He did his best, and his best was excellent. The
public loved him—and bought bonds. However, he never
stopped thinking about his fellow Marines still fighting
island to island in the Pacific:

"After about six months of tours and speeches I found myself doing guard duty at Washington D.C., Navy Yard. I feel like a museum piece. It seemed ages ago since I'd left the South Pacific the previous summer. Washington was a pleasant place. But I wasn't very happy. I wanted to get back to the machine guns. I felt out of things. I've done three years of duty in the Philippines, and it has been my ambition ever since Pearl Harbor to be with the outfit that recaptured Manila. I kept thinking of how awful it would be if some Marines made a landing on Dewey Boulevard on the Manila waterfront and Manila John Basilone wasn't among them."[92]

He was offered a commission, and turned it down: "I'm a plain soldier—I want to stay one." Basilone wasn't a politician or a figurehead: He was a warrior. He refused any Military Occupational Specialty that "didn't involve firing heavy weapons at enemy combatants, telling his commanders that he wasn't going to turn his back on the Marines that needed him."[93]

The bond tour and stateside duty required a different kind of endurance for an NCO like Basilone. Having lived through Guadalcanal once, he now had to live through it, over and over, as he told the stories to media and crowds that wanted all the gritty details. He was reliving the hardest and most gruesome events of his life in what seemed like a nonstop narrative. During the 1943 War Bond Tour, he said, doing a "stateside tour is tougher than fighting Japs."[94]

Basilone demonstrated a different aspect of endurance: mental and emotional toughness. "I was a little shaken, retelling those terrible hours all over again," he said. Yet he still coped, talking at length to journalists and civilians who had no frame of reference for what he had been through.[95] In December 1943, he requested again to return to the war, and this time the request was approved. After a few days' liberty over the Christmas holidays, he left for Camp Pendleton in California for training. Talking about his arrival, Basilone said:

> *"You don't know what a thrill it was to me to walk into one of those battleship gray barracks at Pendleton and see a long line of machine guns parked in the aisle between the bunks. I felt like kissing the heavies on their water jackets."*[96]

Sergeant William Lansford trained under Basilone at Camp Pendleton. He remembers that time fondly. Basilone looking after them like an older brother:

> *"Under the hot California sun, with our faces stuck in the dust of Camp Pendleton, he could pick up a draggy machine gun drill with 'Awright, ya goldbricks. Ya cut the time on settin' them guns up or don't expect no liberty come Friday!' And we did it because we knew he was the best machine gunner in the Corps and we wanted to be like him."*[97]

Lansford was one of Basilone's boys, and the rest of the newly formed 5th Marine Division was envious. He motivated them to look out for each other and pumped up their morale. "His simplicity, his cheerfulness, his grasp of human nature—the charm and easy grace with which he carried his honors—gave us not only confidence but pride."[98]

While stationed at Camp Pendleton, Basilone met his future wife, Lena Mae Riggi, who was a sergeant in the Marine Corps Women's Reserve. "They met at Camp Pendleton. He was very charming, good-looking, yet tough. He was a man of honor and quite a hero. All the ladies thought he was a very good man," said Barbara Garner, a long-time friend of Lena. On July 7, 1944, the couple wed.[99] But their time together would be short: Basilone and the men he was training were heading out to another small Pacific island—one called Iwo Jima. Although he was very happy with his new wife and the life they were building together, the Marine Corps needed him elsewhere. He had already made his decision to go where he could best support his Marines, and he would endure whatever it took along the way.

Basilone exemplified another part of endurance: <u>When in doubt, don't stop. Remember the mission and do the next thing you can do to support it</u>. <u>It's often difficult to know what the next thing is</u>. It's great when you have an organization like the Marine Corps to tell you what the next thing is. Sometimes we find it in ourselves, sometimes from outside authority.

Finding the next thing is based on vision and mission. If you are a part of a larger vision, a mission bigger than yourself, you trust that what you are being asked to do is a part of that mission. The task is not stupid, not arbitrary, but important in a way that you may not yet understand. Small unit leaders often don't completely understand the mission. But they do know whether they trust that their leaders do.

This isn't to say that each Marine doesn't have his or her own mission. Basilone wanted to be with his men, keeping them alive and keeping the battle moving forward. That was his mission. To get there, he still did the things assigned to him, to the best of his ability, even when he didn't especially care for the task at hand. He endured the bond tour and the media, but when the opportunity came for him to present his case for what he saw as his mission, he took it. He explained why he would best serve training Marines to fight and understand the enemy, and leading them onto new battlefields. Because he had been a good soldier and done his duty until that point, his superiors listened to his explanation of his personal mission, and it became their own.

The enduring leader takes a setback and finds new and effective strategies for coping with them in the future. When Basilone had to grab that hot barrel on Guadalcanal he did it unflinchingly, knowing that he'd burn his arms and hands. But he didn't stop there. He looked at the problem and then invented the Basilone Bail, a simple contraption of wire and a handle to put over the barrel to make it possible to move a hot gun intact and on the run. And it's the only way to fire a machine gun from the hip.[100]

And he didn't keep this handy tip to himself. Back at Camp Pendleton, after the bond tour, Basilone worked to train the Marines he eventually would lead on Iwo Jima. He made certain all of his Marines knew about and knew how to use the Basilone Bail, so that none of them would have to lose flesh for want of a glove—or lose their lives because of their inability to move their guns.

Basilone shipped out one month after his wedding to Lena Riggi. When the first waves of Marines went ashore on Iwo Jima, Basilone was there, attached to the 1st Battalion, 27th Marines, 5th Marine Division.

Bronze Star recipient Marine Chuck Tatum hit Iwo Jima as a machinegunner with Basilone. He remembered that:

> *"Steve Evanson, my assistant gunner, and I were in the first wave. We were fighting our way up the steep black sand terraces, hampered by 65 pounds of combat gear. Gasping for breath, we struggled up to the top, where the Japanese gunners could see the whites of our eyes. They opened fire on all the Marines trapped on the black sand beaches of Iwo Jima. They had us in their sights. We were zeroed in on their killing fields.*
>
> *"'I looked back on the beach and only saw one lone Marine standing up. The rest of us were hugging the ground in the prone position. That Marine was GySgt. John Basilone. He was kicking butts and telling Marines to get up and advance or they would surely die on the beach."*[101]

The invasion had ground to a halt, but Basilone's leadership by example got the assault back underway. For Manila John there now was only one objective—the west coast—and he ran for it, his men behind him.[102]

Iwo Jima translates to English as "Sulfur Island." The beaches on Iwo consisted of black sand that stunk of rotten eggs, reminiscent of sulfur. The glassy grains of sand provide next to no traction, and slowed down any movement considerably. Manila John Basilone called to his machine-gunners, "Let's get these guns off the beach." They moved inland as quickly as they could, sinking into sand above their ankles, "but Manila John would never see Dewey Boulevard again."[103] He was twenty-nine years old.

Tatum continued fighting. He recalled:

> *"On the tenth day we didn't even own half of it yet. The 'landlords' held more of it than we did. As I reflect back on it ... I think Iwo Jima was more than a battle. It was a thirty-six day descent into hell on earth ... an apocalypse in the Pacific."[104]*

On her thirty-second birthday, Lena learned that John had been killed in action when she received a telegram:

DEEPLY REGRET TO INFORM YOU THAT YOUR HUSBAND, GUNNERY SERGEANT JOHN BASILONE, USMC, WAS KILLED IN ACTION FEBRUARY 19, 1945 AT IWO JIMA, VOLCANO ISLANDS, IN THE PERFORMANCE OF HIS DUTY AND SERVICE TO HIS COUNTRY. WHEN INFORMATION IS RECEIVED REGARDING BURIAL, YOU WILL BE NOTIFIED. PLEASE ACCEPT MY HEARTFELT SYMPATHY.

The telegram was signed by General Alexander Vandegrift, who had been with Basilone at Guadalcanal, and who had presented the Medal of Honor to Basilone in Australia.

Basilone also was posthumously awarded the Navy Cross. His citation reads in part:

For extraordinary heroism while serving as a Leader of a Machine-Gun Section, Company C, 1st Battalion, 27th Marines, 5th Marine Division, in action against enemy Japanese forces on Iwo Jima in the Volcano Islands, 19 February 1945. Shrewdly gauging the tactical situation shortly after landing when his company's advance was held up by the concentrated fire of a heavily fortified Japanese blockhouse, Gunnery Sergeant Basilone boldly defied the smashing bombardment of heavy caliber fire to work his way around the flank and up to a position directly on top of the blockhouse and then, attacking with grenades and demolitions, single handedly destroyed the entire hostile strong point and its defending garrison. Consistently daring and aggressive as he fought his way over the battle-torn beach and up the sloping, gun-studded terraces toward Airfield Number 1, he repeatedly exposed himself to the blasting fury of exploding shells and later in the day coolly proceeded to the aid of a friendly tank which had been trapped in an enemy mine field under intense mortar and artillery barrages, skillfully guiding the heavy vehicle over the hazardous terrain to safety, despite the overwhelming volume of hostile fire. In the forefront of the assault at all times, he pushed forward with dauntless courage and iron determination until, moving upon the edge of the airfield, he fell, instantly killed by a bursting mortar shell. Stouthearted and indomitable, Gunnery Sergeant Basilone, by his intrepid initiative, outstanding skill, and valiant spirit of self-sacrifice in the face of the fanatic opposition, contributed materially to the advance of his company during the early critical period of the assault, and his unwavering devo-

*tion to duty throughout the bitter conflict was an inspira-
tion to his comrades and reflects the highest credit upon
Gunnery Sergeant Basilone and the United States Naval
Service. He gallantly gave his life in the service of his
country.* [105]

Basilone's fate on Iwo Jima may not seem a happy ending, but it is one true to his own sense of self and duty. He
understood that victory was an outcome of a process that
took fear and uncertainty and distilled it into focus on
smaller goals and actions to reach those goals. He knew
that leaders, perhaps most especially small-unit leaders,
make meaning for followers. In times of trouble, subordinates will look to their leaders for guidance. A terrific
leader helps them to understand the problem, know that a
plan exists to address it, and know that they are part of the
solution.

Basilone was given mission-type orders and then was
left alone (intentionally or not) to make them happen. He
delivered. He also trusted his men to do what he asked
them to do. He knew their training and abilities, that they
could in fact do the job. He trusted that in the crunch, they
actually would do it. And his men trusted him, as he was
right there alongside them. He asked nothing of them that
he was unwilling to do himself. By exemplifying his ability
and willingness to endure what was required in impossible
conditions, his fellow Marines trusted that what he was
asking of them was in fact possible—and necessary.

They trusted him to make decisions that minimized risk
where possible, and he trusted that they would carry out
their orders—despite any hazards—for the overall benefit
of the unit.

The enduring leader defaults to responsibility. If something must be done, then it must be done even if the best resources or relevant training aren't available. Basilone continually took immediate ownership and accepted the challenges presented to him. He understood the consequence of failure. He had a bias for action. When in doubt, he did the best he could with what he had. Frequently, hesitation while waiting for the perfect circumstances means no decision at all, a quick path for leaders to lose credibility. Basilone understood that.

The incredible endurance shown by Sergeant John Basilone, at war where he excelled and at home where he persisted, took strength of body and of willpower. Some would say strength of mind far outweighs physical capabilities. We'll ponder that next.

Part 2

Mind

Chapter 6

Knowledge

The Nation that makes a great distinction between its scholars and its warriors will have its thinking done by cowards and its fighting done by fools.
—Thucydides (460 BC–395 BC)

There's little in a Marine NCO's skill set that beats intimate knowledge of how things work. It might be an ability to weave in and out of a tactical problem or it might be admirable skill with a particular weapon, but the business of knowing what to do and how to do it lifts the leader above the crowd. Marine Corps history is loaded with examples but one of the most colorful and instructive involves Lou Diamond, arguably one of the best mortarmen and the saltiest NCO the Corps ever tolerated.

Knowledge goes beyond the facts of the leader's job. It is also knowledge of your team: who they are and what motivates them. It is knowledge of the culture in which you work, so that you understand what your superiors' goals and missions are. And it also is self-knowledge: unflinchingly knowing your own strengths and weaknesses, and having a desire to excel.

Master Gunnery Sergeant Lou Diamond exemplified knowledge and its application to his work, and to the Marine Corps. Diamond learned early in his career that if

you're really, really good at what you do, you get to be
weird.

The Marine Corps has been accused of putting out
cookie-cutter warriors, interchangeable parts in the engine
of war. Even though our popular culture is full of images of
gnarly drill instructors and interchangeable recruits, the
Marine Corps is among the most open-minded and
knowledge-oriented organizations in the world.[106] The
Marine Corps is filled with colorful characters, unique
personalities, and individualists. And Diamond is the gold
standard in that measurement.

Master Gunnery Sergeant Leland
"Lou" Diamond, who was on many
occasions decorated for bravery, was
one of the most famous of all the
fighting Leathernecks. The
Diamond legend lives on in Marine
Corps tradition and history. (Public
domain.)

On the surface, Diamond
never looked quite like the
stereotypical Marine. His stern face with its twinkling eyes
and a nose shaped by several fists bore a jaunty white
goatee. In garrison, he often surrounded himself with pets
that were as cantankerous as he was—and about as attrac-
tive. In 1918, one of his friends, Master Gunnery Sergeant
Michael T. Finn, ran in to Diamond with one of his pets:

*"One day, coming back from Nicaragua, I got off the train
at Quantico…and there was Lou Diamond with his*

*bulldog, Bozo. This Bozo was the ugliest bulldog I ever
saw. But, I would say that Bozo was considerably prettier
than Diamond."*[107]

In addition to Bozo, Diamond kept many other pets,
which he acquired at various posts and stations leading up
to the start of World War II. At the time he was preparing
to leave for the South Pacific in 1941, he owned a particularly cranky goat named "Rufus" as well as a couple of
chickens whose names are better left unmentioned.[108]
These were left in the care of a farmer during Diamond's
deployment with the 1st Marine Division heading out for
the first allied offensive in the Pacific. While he was in the
Solomon Islands, Diamond read reports of meat shortages
back in the States with considerable alarm for the potential
fate of his four-legged friends. When *Leatherneck* magazine
was questioned about the whereabouts of Diamond's
animals, they replied in the January 1944 issue, "For all we
know 'Gunny' Lou Diamond has pets scattered all the way
from New River to Guadalcanal."

Leland "Lou" Diamond was born May 30, 1890 in Bed-
ford, Ohio. Maybe. Diamond had to stretch the truth a bit
about his age when he first enlisted in the Corps in 1917.[109]
Then, as now, the Marine Corps preferred to recruit young
men for its combat troops. Diamond's friends claimed his
supposed birth date came from someone else's tombstone,
according to Keith Milks writing in the *Marine Corps
Times*. He was at least twenty-seven years old and working
as a railroad switchman when he enlisted in the Marines on
July 25, 1917 in Detroit, Michigan.

In some ways, running a marshalling yard prepared Diamond for combat duty with the Marines. The yardmaster

had to be an expert at his job. He had to be quick and smart, and lucky, too. If not, he could easily lose an arm or a leg, or everything. The yardmaster before him as yard-master made one bad move and was crushed to death in a coupling accident. It was noisy, grinding, gritty and dangerous work, and Diamond was probably glad to leave and enlist for the duration of World War I.

"I am going to fight for my country," he said, and he never stopped battling.[110]

Master Gunnery Sergeant Finn remembered Diamond well. He told his commanding officer, Captain Alfred H. Noble, "This Diamond fellow ought to make corporal for he is the hardest-working son-of-a-gun in the company, even if he is, probably, old enough to be outside of the draft."[111]

As a corporal in January 1918, Diamond shipped out for duty in France and saw action with the 6th Marines at Chateau Thierry, Belleau Wood and the Meuse-Argonne. He was promoted to sergeant and marched to the Rhine with the Army of Occupation.[112] By war's end he was beginning to be known as "Mr. Leatherneck" for his devotion to the Corps. He returned to the United States in 1919 and accepted a discharge, although that seemed out of character for the dedicated Marine.

But after returning to the rail yards, Diamond just couldn't make the civilian life work for him. He reenlisted in the Marine Corps in September 1921. As a combat veteran, promotion came quickly, and he had his sergeant's stripes back by 1925, when he began work as an assistant armorer at Parris Island. That job also grew tedious, and he shipped out with the 4th Marines to Shanghai. During

World War I, many western powers removed their troops from China in order to fight at home. After forces from Britain, France, Austria-Hungary, Germany, and Russia left China in 1914, Japan declared war on Germany and grabbed German territory throughout the Pacific and in China.[113] The 4th Marines, including Diamond, were needed to act as armed guards for privately owned American boats and some U.S. Navy gunboats on the Yangtze River as it was filled with bandits robbing watercraft with abandon. This conflict, according to Diamond, was "not much of a war," and so he returned to the United States, arriving at Mare Island, California on June 10, 1933. By then he was a gunnery sergeant.[114]

When his old outfit was ordered back to China, Diamond rejoined the 4th Marines and shipped overseas in June 1934. He was transferred to the 2nd Marines in December 1934 and returned to the States in 1937. Two years later, he was a Master Gunnery Sergeant serving at the Depot of Supply in Philadelphia, using his vast field experience to help design a new combat pack for Marines.

Following the Japanese attack on Pearl Harbor in December 1941, Diamond shipped out for Guadalcanal with Company H, 2nd Battalion, 5th Marines as the noncommissioned officer in charge of the unit's mortars. Although he was fifty-two years old by that time, he was such a demonstrated expert with those tubes that there was never a question of not taking him into combat.

In fact, Diamond had become something of a living legend. Among many fables surrounding his service on Guadalcanal, the most familiar is the claim that Diamond once dropped an 81mm mortar round directly down the

stack of a Japanese warship steaming down the channel
between the northern shores of New Georgia and Guadalcanal and the south coast of Choiseul and Santa Isabel
islands known as "The Slot." Although today that story has
been deemed extremely unlikely, nearly all observers credit
him with driving the enemy ship from its firing position
with harassing fire and several near misses.

Another story involved Second Lieutenant Charles A.
Rigaud, who was in charge of the mortar platoon of a heavy
weapons company at Quantico. On maneuvers at Fort
Indiantown Gap in Lebanon County, Pennsylvania, the
men were practicing firing mortars over an abandoned
farmhouse. Diamond bragged that he could drop a shell
right down the chimney. The first shell fell short, but the
second produced a huge plume of smoke and dust right out
of the chimney.[115] It is possible that he hit the basement
and just sent debris up the chimney—but it's also possible
that he was just as good a shot as he thought he was.

War correspondent John Hersey was embedded with
Diamond's team on Guadalcanal in early October, 1942.
He wrote:

> *"I heard a mortar battery making twice the noise a bat-
> tery usually makes … the extra noise was shouting … it
> was Master Gunnery Sergeant Lou (Leland) Diamond,
> who was said to be approximately two hundred years old.
> Presently I saw him—a giant with a full grey beard, an
> admirable paunch, and the bearing of a man daring you
> to insult him.*
>
> > *"They told me that Lou was so old that there was some
> > question whether to take him along on such a hazardous
> > job as the Solomons campaign … Here he was, proving*

that even if he out Methuselahed Methuselah, he would
still be the best damn mortar man in the Marines."[116]

Today, mortars are still a vital part of modern warfare.
Modern mortars are precision weapons of high accuracy,
capable of rapid, highly effective fire.[117] The high, curved
trajectory of the shell makes it possible to emplace mortars
in positions where they can be protected from weapons that
fire with a flatter trajectory. It's like lobbing snowballs over
a wall to hit your friends—they can't see you or hit you
back with a straight throw. The mortar is simple and not
particularly prone to stoppages.

Despite this simplicity, mortar gunnery requires difficult
computation to ensure near-first-round accuracy. Although
it is an area-fire weapon, modern infantrymen expect the
mortar man to place the round on target in a hurry. In
Diamond's heyday mortarmen had to make difficult
computations using a plotting board to find the gunnery
solution, where firing data are applied to the ammunition
and the mortar so that the fired projectile bursts at the
desired location. Those data are based on details such as
"direction, horizontal range, and vertical interval from the
mortar to the target, the pattern of bursts desired at the
target," and the weather.[118]

Diamond was the best there was. He made magic with
the plotting board. The Marines around him admired his
ability to calculate the gunnery solution in his head. If he
could see the target, he knew the range, the charge, the
deflection, and the direction, and could then call it down to
the guns without any additional computation. By the time
he landed on Guadalcanal, Diamond knew the mortar
intimately, having studied it for more than twenty years.[119]

General A.A. Vandegrift, who at the time was in command of the 1st Marine Division, issued a citation attesting to the Master Gunny's value:

"To every man in your company you were a counselor, an arbiter of disputes, and an ideal Marine. Your matchless loyalty and love of the Marine Corps and all it stands for, are known to hundreds of officers and men of this Division, and will serve as an inspiration to them on all the battlefields on which this Division may in the future be engaged."[120]

Though not a "spit-and-polish" Marine, Diamond proved himself such expert with both 60- and 81mm mortars, his accurate fire was credited as the "turning point of many an engagement in the Pacific during World War II."[121]

<u>Diamond never stopped learning.</u>

<u>Once a leader has knowledge like Diamond had, it is an obligation to share and teach others so that the knowledge becomes a unit asset.</u> Marine Corps Doctrinal Publication 6 notes that, "We focus on the value and timeliness of information, rather than the amount, and on getting that information to the right people in the right form."[122] One way of sharing that knowledge is through after-action debriefs. A leader won't always know what his or her team learned from a mission until they are asked. Debriefing after both success and failures allows for lessons learned to be passed on to others who can use that knowledge. When debriefs are performed consistently with fairness and critical review, team members share openly, knowing that they will not be punished for failure, but rather that all will

learn and develop their own skills and the skills of the team. That was as true in Diamond's time as it is today.

In 2010, members of Alpha Company, 1st Battalion, 23rd Marines, including a 60mm mortar team, practiced live-fire exercises at Camp Pendleton, California, in preparation for deployment to Afghanistan. After the exercises were completed, the Marines were able to debrief and share with one another what they had learned, both what worked well, and what needed improvement. Lance Corporal Bobby Henrichsen, one of the Marines involved in that exercise, explained how training and sharing knowledge helps the entire team: "This type of training helps build trust with all the other Marines in the platoon. The platoon already has good chemistry, but platoon attacks help us work together as one unit while building communication."[123]

Modern technology makes sharing knowledge easier and improves accessibility. One example is the communities of practice, in which individuals with similar functions can share knowledge with each other through online bulletin boards and forums. Too much reliance on technology can be hazardous, however. Use it where you can, but don't become over-dependent on it.

Sharing knowledge with subordinates can feel to some leaders like they are giving up control, and they may be loathe to do so. In reality, though, leaders are not effective because they are the knowledge holders. Rather, the best leaders are the ones who make knowledge available to their teams, and understand how best to deploy that knowledge in the best possible manner. In this modern age, knowledge can have a remarkably short shelf life. As the battlefield—

whether in Iraq or Afghanistan or a corporate office—
changes constantly, the rapid transfer of knowledge becomes vital. Diamond knew that long before the age of the
internet.

As World War II progressed, Diamond's health and age
precluded any further combat service in the South Pacific,
so in 1943 Diamond was transferred back to the States
where he became a premiere instructor at Parris Island and
Camp Lejeune. He taught combat-bound Marines more
about mortars than anyone thought there was to learn. He
was rough-and-tumble and quite physical when young
Marines didn't jump at his notorious bark. But he also
saved a lot of lives by making those young Marines tough
enough to survive combat.

Diamond's duties allowed him to interact with officers
as well as enlisted men. In July of 1944, he was running a
boot platoon through the decontamination chamber. He
made it clear that knowledge meant a lot more than just an
education:

> *"With mattresses above our shorn heads we were under-*
> *standably awed by his presence. There wasn't room for*
> *another hashmark on his sleeve, and he was already a*
> *legend within the Marine Corps. He planted his booted*
> *right foot into our rear ends with professional aplomb and*
> *with favor toward none. He knew we were members of*
> *that elite group later referred to as OC's [Officer Candi-*
> *dates], but he was unimpressed. Our nervousness had an*
> *adverse effect upon our efficiency which Lou Diamond*
> *noted with obvious satisfaction and took advantage of,*
> *calling us 'college boys, lots of education, but no damn*

sense!' and, 'Boy, I bet you come from an agriculture col-
lege!'"124

Diamond rejected several opportunities to apply for a commission, explaining that "nobody can make a gentleman out of me."

In 1943, *Time* magazine wrote about Diamond's work at Parris Island:

"Master Gunnery Sergeant Leland Stanford Diamond, the Marine's Marine and the Corp's most famed mortar expert, is back from the wars....But Parris Island is as good a backdrop as any for Diamond's fabulous personality. In his new job he greets recruits with a wounded-bull roar, shoves them through an assembly line of showers, haircuts, lice inspection and clothing issues at the rate of 200 an hour. He bawls countless lectures at awed recruits (and quaking second lieutenants), lectures he has learned by heart in 26 years of professional soldiering.... And each night there are two to three cases of cool beer on hand to be shared with a veteran M.P. sergeant who can almost keep up with Lou at cribbage.

"Last week Lou Diamond's peaceful, foamy world was threatened: the post-exchange staff was taking off its first day in months for a picnic, and no beer would be sold at Parris Island for a whole day. Lou's roars of outraged thirst reverberated through the battalions, reached the post exchange just in time. Eventually the picnic went off swimmingly with endless free beer. Special last-minute guest: Master Gunnery Sergeant Diamond."125

Diamond wasn't going to let this report stand unchallenged. He wrote back to *Time*, and his letter was published on January 10, 1944:

> *Mortarman Diamond, U.S.M.C.*
> *Sirs:*
> *Received the magazine and will put you straight on that story (TIME, Dec. 13).*
> *First, I have no middle name.*
> *Second, I did not go to the party at all....*
> *Now for the good of those poor boys that went six months without a party:*
> *They get paid extra for working in the canteen and they are not out in the rain. They will never have to do combat duty....I think someone ought to buy them a nursing bottle for crying to the public about their party.*
> *This outfit I have works just as long and harder than they do and gets not a cent extra for it, and has no time for a beer party till after work.*
> *Lou*
> *Parris Island, S.C.*
> *P.S. I do not play cards with M.P. Sergeants.*[126]

Whether or not his target was M.P. Sergeants, Diamond did love to play cards. In fact, he loved many kinds of mental challenges, and he had a vast curiosity about the world. And curiosity is one of the best motivators to turn information into knowledge. Curiosity keeps the mind in an active mode, asking questions about the information it gathers. These questions open the mind to new ideas, and when new information arises, it looks for patterns and applications to the knowledge it already has.

Michael Stokey, who served as a sergeant in Vietnam, has a lot in common with Diamond. He, too, was no 'spit-and-polish' Marine. He also had an insatiable curiosity and great knowledge:

> *"I knew some things in combat, even though I hadn't studied the proper terms for everything. I wasn't a student of formal tactics. Terrain and technology were the most important things to me, when analyzing a situation. And it's applicable throughout time. When I learned about the Macedonian phalanx back in the time of Alexander the Great, I knew without study that a phalanx can't work in a jungle. A large group of men with long spears and locked shields might be great on a flat plain, but it's useless in a jungle."*[127]

Diamond was curious about everything, and his passion was accuracy. He liked to recall a baseball game at Tientsin in 1934, "when a Marine batter hit a line drive that killed a sparrow in flight. In this accident he sees a higher goal for precision marksmen."[128]

Today's mortarmen still must master the knowledge of their weapon, just as they did in Diamond's time. On December 10, 2010, Marines at Camp Pendleton, California, practiced during an 81mm mortar shoot. According to a story from the 1st Marine Division, their unit was set to deploy to Afghanistan in 2011, and in preparation, this Mortar Platoon used the time to develop new mortar teams and to cross-train with their 60mm counterparts.

The Marine Corps has recently acquired a new 120mm mortar. The 81mm team must now learn the new system, and the 60mm sections must master the 81s.

"This shoot allows us to know our weapon inside and out. We could do notional firing and practice our guns all day," said PFC Joseph A. Callas, a mortar man with the 81mm Platoon. "Until we get to shoot rounds out of it, we are not going to know how the gun will feel in a combat situation."

"The quicker we get them out here and to these mortar shoots, the quicker they learn and the cohesion builds." said 1st Lt. Patrick B. Oshea, the 81 mm Mortar Platoon commander. "It's very important to get everyone you are going to deploy with out here because the unit gains its strength out here in the field."[129]

Cross-training applies beyond different types of the same weapons. Marine training aspires to have every Marine understand all the elements of Marine warfighting, regardless of his or her military occupational specialty. This kind of cross-training helps Marine small-unit leaders understand their own part of the mission, and how it fits into the bigger picture.[130] And, by understanding the work that other Marines are doing, their sense of shared culture is reinforced.

The cohesiveness that binds Marines together comes, in part, from this shared knowledge. Marines expend enormous effort in learning and teaching their own history from the first day of recruit training. Through this study, recruits learn about the origins and history of the Marines, how the Marine Corps values were developed, and what is expected of them as Marines, should they earn the title. And they learn why they fight, not just how to fight.

Marine Small Unit Leaders are expected to continue to expand their knowledge through Enlisted Professional Military Education (EPME). Traditionally, EPME has been limited to individual courses at various enlisted academies or distance learning courses through the Marine Corps Institute. Today, the Marine Corps is providing a more career-long approach. The goal is to ensure that corporals and sergeants, for example, are capable of applying tactical and technical skills appropriate to their levels of responsibility in real-world situations, using a combination of web-based technology and resident instruction.[131] In addition to teaching specific and necessary skills, young NCOs also learn the habit of life-long acquisition of knowledge.

Cross-training for knowledge acquisition applies to civilian life as well. Modern careers are not like they once were. The old expectation of staying with one firm for a career lifetime, moving us as positions become available, is no longer the norm, or even common. Now, professionals move from corporation to corporation. A potential leader who limits his or her knowledge to one company is also limiting his or her own career, as well as the ability to lead elsewhere. Even if a professional stays with one firm, there is so much diversity in modern organizations that the leader still needs to continually update his or her knowledge bank.

Gunnery Sergeant Bruce Whitfield, who served in both Desert Storm and Desert Shield, has served the Corps in a variety of duties. One of his most rewarding tours was at 1st Marine Division Schools as the Chief Instructor for mortars. His course curriculum included mortar gunnery, forward observation, and fire direction center at the intermediate level. "When I took over the course it was in disarray," he said. "All the classes had to be written and school books had to be printed in one month. Luckily, I had a group of outstanding NCOs." Sergeant Williams and Sergeant Paulsen taught forward observers, Sergeant Izzo led the gun line, and Sergeant Garret led the fire direction center. Each of Whitfield's NCOs was a combat veteran who had fought in Falluljah and Ramadi in Iraq. These Marines were eager to teach and share their hard-learned knowledge of mortars in urban combat with young Marines in the 1st Marine Division, constantly researching and improving the material that they were using to teach Marines.

Whitfield would invite combat veterans from previous conflicts to talk with his students. One of these speakers was Chief Warrant Officer Gilbert Bolton, a Vietnam veteran who was awarded a Silver Star for heroism in combat:

> *"He told us a story of Corporal Amendola, who was in his 60mm mortar section. He also earned a medal that night, when their gun position was being overrun. Corporal Amendola was last seen alive fighting in his mortar pit swinging the mortar barrel like a club at the NVA until he was finally killed. He was awarded the Navy Cross posthumously."[132]*

After Gunner Bolton's story, Whitfield talked with his NCOs about naming our mortar building at Camp Margarita aboard Camp Pendleton after Corporal Amendola.

They all agreed it would be an honor. These Marines took on the task of renovating the entire building. Whitfield arranged a dedication ceremony with Corporal Amendola's family in attendance, along with the former Platoon Sergeant, Gunner Bolton.

A lot of salty mortar men talk about Lou Diamond, and Whitfield and his Marines were no exception. As a mortar instructor, Whitfield would discuss Diamond with his students. "Many of the new Marines looked at me with a puzzled look, because they had no idea who I was talking about," Whitfield said in an interview. "That's why I continued to talk about him to each class in order to continue his legacy." Diamond's picture was displayed prominently in the classroom, and many of the curious students asked about the man with the goatee. The NCO Instructors would tell his story, including some of the more fantastic. Whitfield continued:

> *"He was most famous for his mortar gunnery skills and incredible accuracy in chasing off a Japanese Cruiser during World War II; some stories even say he fired a round that went down the smoke stack. I guess we will never know and it doesn't really matter. He's a Marine hero, and most important, a mortar man."*[133]

Master Gunnery Sergeant Lou Diamond retired on November 23, 1945 and returned to Toledo, Ohio. But when the Korean War broke out five years later, he responded by marching over to the local Marine recruiter and volunteering to serve.

He was politely rejected because of age. So he wrote the Commandant of the Corps, requesting a call to duty. General Clifton Cates graciously turned him down, but added that if the war got worse, he would keep the old mortarman in mind[134]. He knew that, if needed, Diamond would be ready and able to share a vast store of knowledge.

Knowledge is more than information. Memorizing appropriate facts and figures is a great way to get started on the path to knowledge, but it's just the first step. In today's world, information is everywhere, and a good leader is adept at retrieving information when needed. Because data are so omnipresent, what becomes more important is an ability to discern the useful information from the superfluous. Information by itself is not going to help leaders make decisions unless they understand how to apply that data to the situation at hand. Diamond knew that, and so do today's NCOs.

As we pass through the leadership aspects of the mind, we're building towards action—making excellent decisions and following through with them in a timely manner. Before you can begin to make decision, though, you need to know what your options might be. And these options are found through knowledge. Of course, the practical aspects of having knowledge demand that a leader use that knowledge to make sound decisions. That involves the leadership attribute of judgment.

Chapter 7

Judgment

*No victory is possible unless the commander be energetic,
eager for responsibilities and bold undertakings....unless
he be capable of exerting a personal action composed of will,
judgment, and freedom of mind in the midst of danger.*
–Ferdinand Foch, Marshal of France (1851–1929)

In most cases, judgment is simply a matter of making a choice. When there are options—and there usually are in most human endeavors—we process the upside and downside, assess the potential outcomes and choose a course of action. Personal and professional lives involve thousands of choices on a daily basis. Choices may involve judgment in something as basic as what to eat or what to wear: the simple stuff that we do almost instinctively. And then there are situations that carry much more weight and involve serious consequences for bad choices. Those are the sort of mental and moral choices that cannot be made purely on instinct. In fact, when lives are on the line, proper judgment is often in direct conflict with human instincts.

For leaders like Marine NCOs there's a lot more involved and a lot more at risk, especially on battlefields where a lapse in judgment—especially on the battlefield—can cause a vital mission to fail or cost casualties. Marine NCOs have been required to exercise sound judgment in stressful situations throughout the history of the Corps and for the most part their record is superlative. That legacy, demonstrated for centuries on far-flung worldwide battle-fields for decades, lives on among the young corporals and sergeants leading from the front in the current conflicts in the Middle East. It is instructive to examine their experi-ences in situations in which decision-making and the information required to exercise proper judgment often involves cultural imperatives alien to their background and frequently shoves responsibility down to the lowest levels.

In places like Kuwait, Iraq and Afghanistan, tough judgments involving serious consequences often are made by men and women wearing chevrons rather than stars. It is the nature of what's called "asymmetrical warfare," in which conventional forces do battle with irregular formations in areas crowded with often hostile populations. This kind of fight is what the Marine Corps has dubbed the "Three Block War" and it involves young NCOs making signifi-cant decisions on their own with little or no input from senior commanders. Today in Afghanistan, for instance, small units—squads and platoons—often operate in areas that are beyond the immediate influence of higher com-mands. Young Marine NCOs are required to make deci-sions that in previous conflicts would have been bounced up to a senior officer for guidance.

General Charles Krulak, a former Commandant of the Marine Corps, wrote extensively about the *Three Block War* and its impact on the need for NCOs to display good judgment. He has noted that the lessons of:

"... *recent operations, whether humanitarian assistance, peace-keeping, or traditional warfighting, is that their outcome may hinge on decisions made by small unit lead-*

*ers, and by actions taken at the lowest level....today's
Marines will often operate far "from the flagpole" without the direct supervision of senior leadership....In order
to succeed under such demanding conditions they will
require unwavering maturity, judgment, and strength of
character. Most importantly, these missions will require
them   to   confidently   make   well-reasoned
and independent decisions   under   extreme   stress—
decisions that will likely be subject to the harsh scrutiny of
both the media and the court of public opinion."[135]*

Nothing in the current Middle East experience among
Marine NCOs demonstrates the wisdom of Krulak's
assessment better than the constant interaction with
civilians experienced by patrols often led by corporals or
sergeants throughout Afghanistan's rugged countryside. In
2004, Zabul province in Afghanistan was considered fertile
ground for insurgents. Among the risk factors were a small
population, an insecure border with Pakistan, and little
central authority. Many residents had no contact or information about the world outside of their own village or
town. Today, however, Zabul is a rare success in Afghanistan's turbulent southern region.

Success is a relative term in this troubled region, and
insurgents still remain in the province, living in some of the
country's worst poverty. However, considerable transformation has occurred since 2004. At that time, U.S. troops
began a large-scale push into the region. Some residents
were isolated enough to believe that the incoming military
was Russian[136] forces returning to the area for another
attempt at occupation. In reality those forces were primarily

U.S. Marines moving into an area of Afghanistan that was considered fertile ground for insurgents.

Some of the success in the region was the result of civic efforts such as such American-funded building projects and the solar-powered street lights installed in 2009 that allow shops to remain open longer and the streets to be safer while not taxing the extremely fragile power grid that powers Afghani homes.[137] Other progress in the area can be attributed directly to the efforts of Marines who have patrolled the area for years now, engaging Taliban insurgents to keep them at bay, as well as engaging with the people of the province to build trust and security. However, it often is difficult to judge what investments will give the greatest rewards for both groups. In order to make judgments on what projects to fund, and what the local civilians need, female service members are stepping in.

Recognizing that making contact with Afghan females in Afghanistan was virtually impossible for male Marines, planners came up with an innovative concept involving specially-trained units of women Marines organized into Female Engagement Teams (FETs). Although Afghani women are barred from talking to male strangers, there is nothing in the cultural norm that says they can't associate with other females.

The Marine Corps took rapid, effective advantage of that and officially launched the FET program on October 1, 2010, although some teams operated earlier. Female Marines are allowed entrance into the homes of Afghani women, who often are more open to conversation and to sharing what their hopes are for their piece of Afghanistan. These NCOs found themselves functioning as reliable bridge-builders between cultures and valuable sources of gut-level intelligence gleaned from Afghan women who shared their visions and insights with the American women.

U.S. Marines serving with the female engagement team (FET) attached to Golf Company, 2nd Battalion, 9th Marine Regiment visit a practicing midwife in Marjah, Afghanistan, Dec. 31, 2010. FET members met with her in hopes of starting prenatal classes for women in the area. (U.S. Marine Corps photo by Cpl. Marionne T. Mangrum/Released.)

"By utilizing our female troops, we have an opportunity to greatly expand the portion of the population with which we engage," said James Judge, a NATO spokesman at the International Security Assistance Force Joint Command.[138] His assessment comes from the realization on the part of coalition planners—arrived at through hard experience— that while Afghan females in a tribal society may appear to be powerless and strictly background players, they actually are quite influential within their families and thus have a dramatic impact on decision-making. And they can often provide a very accurate measure of what's working and

what's counter-productive in efforts to influence populations caught in the middle of the struggle between the
government in Afghanistan and the Taliban forces who
want to destroy its influence.

The teams generally are led and directed by young
NCOs who are called on to make serious judgments based
on military, cultural and gender concerns on every mission.
For Marines like Sergeant Francini A. Fonseca, who served
with one of the Marine Corps' FETs in Afghanistan in
2009, it was a character-building experience she'll never
forget. It's also one that seems unlikely for a young woman
from Brazil who took a long, circuitous route into the
badlands of Afghanistan.

Fonseca was born and raised in Brazil and came to the
United States when she was nineteen years old. In November 2006, she moved from Brazil to Boston looking for
better opportunities for success. With an associate's degree
in computer and software science, she worked as a computer technician in Brazil, but lost her job when economic
hardship hit the country. High-tech jobs were hard to find
so she worked in a variety of tasks to support herself and
lived with her father who had emigrated to America ahead
of her. Along the way she became a U.S. citizen following a
line of family members who had been coming to America
over nearly half a century seeking better lives and increased
opportunities.

"I am really proud of being a United States citizen,"
Fonseca said, "and would fight any war for the nation that
supports me and my family so well." In April 2007, she
decided to put action to those sentiments and joined the

U.S. Marine Corps. Like all female candidates, she attended recruit training at Parris Island where she excelled, graduated as a squad leader and was promoted to Private First Class. She trained as an aviation electronics technician and served at various commands supporting Marine training and repairing vital equipment but the lure of doing something more active and direct to support American efforts in the Middle East was strong. Fonseca wanted in on the action. She volunteered for service overseas and found herself in Afghanistan in the fall of 2009 where she first heard about the Female Engagement Teams. It didn't take her long to volunteer.

In the male-dominated arena of infantry operations, there was some resistance to females working on the firing line. Fonseca said that:

> *"The team had to face male Marines that think female Marines cannot hang in combat and actually perform the duties successfully. Well, I have always had one thing in my mind: No matter what people think and judge you for, just do your best and give 100 percent at all times, letting your actions speak for you. I was focused on the mission. What excited me the most is the fact that FET allowed me to help the Afghani community and combat the Taliban at the same time."[139]*

As a team leader in the program from November 2009 to April 2010, Fonseca was pivotal in leading female Marines on missions that involved everything from vital intelligence gathering to physical security efforts in searching Afghan females for hidden weapons and explosives to general life improvement for the Afghani population. On every one of these missions she had to make judgments and decisions. Was what she was hearing from Afghan females reliable or was it idle gossip? Was what she discovered the real deal or just disinformation designed to mislead the Americans operating in the area? There were few reliable yardsticks in Fonseca's efforts. Her judgments were made based on a combination of military experience, cultural understanding and an ability to read female motivations and psychological signals.

Her FET and others working in the Afghan badlands provided vital insights that guided Marines in making tactical decisions and often helped them avoid deadly improvised explosive devices (IEDs). Some Afghan females knew where these killers were planted and often wanted them removed or eliminated to avoid the bloodshed that followed detonation. Most of these women had a vested and personal interest in keeping their families and their children safe and did not mind sharing valuable information with the Marines as long as it was just sort of woman-to-woman conversation that would not lead to repercussions from the men who planted those devices. That concern led Fonseca and her female team members to consistent judgments involving a delicate balancing act to protect their sources.

That judgment and a generally superior performance as an FET leader in Operation *Enduring Freedom* brought Fonseca a Navy and Marine Corps Achievement Medal. Her team conducted more than two hundred engagements with the villagers. They also launched fifteen medical outreach clinics directly aiding more than three hundred families. The team facilitated economic development, augmented access to education, and improved health care. Fonseca's FET was also credited with providing detailed information about Taliban movements and the insurgent cells making IEDs.

The necessary judgments that arise in places like Afghanistan when dealing with local civilian populations frequently land on the shoulders of squad leaders and other small-unit leaders like Fonseca. They are required to make good judgments themselves, and also to monitor the decisions made by their subordinates. Often, these judgments can have important implications, such as judging whether a situation requires diplomacy or force. In asymmetrical warfare, smaller units are often relied upon more frequently than during conventional campaigns. Although a region could be subdued by sheer force and large numbers, this kind of heavy-handed strategy often leaves the people who live there very unhappy, creating an environment that encourages the insurgents to move right back in once allied armed forces move on. David H. Freedman, author of *Corps Business: The 30 Management Principles of the U.S. Marines*, said: "Small units are the core of anti-insurgency tactics stressing optimum force size—the smallest footprint with the biggest impact…"[140] As leaders of these small units, sergeants and corporals spend their time not only fighting, but learning about local traditions, customs, and the personal lives of the locals who live in their area of operation.

The Marine Corps has struggled to develop guidance for young corporals and sergeants, but so many of the judgments these young men and women must make are situational. No template fits all situations. To assist these NCOs, the Marine Corps created the *Small Wars Manual*, a comprehensive guide for conducting counterinsurgency operations. Although this doctrine was first published in 1936 and revised in 1940 following Marine involvement in Central America, it is still considered today a valuable guideline for conducting operations in Afghanistan and Iraq.[141] Among other things, the manual discusses appropriate behavior and conduct of Marines when dealing with civilian populations during combat operations short of declared, conventional war. It has particular meaning for young Marines who will encounter populations that may have no idea who Americans are or what motivations guide their behaviors.

> *"They judge the United States and the ideals and standards of its people by the conduct of its representatives. It may be no more than a passing patrol whose deportment or language is judged, or it may be fairness in the purchase of a bunch of bananas."[142]*

Whether a bunch of bananas in the mountains of Nicaragua or a serving of flat-bread in Afghanistan, how Marines deal with the people offering or selling is crucial. The key is that local populations coming into contact with Americans for the first time or only rarely in their experience must believe the odd foreigners are not there to steal or exploit despite their power to do so.

> *"There is no service which calls for greater exercise of judgment, persistency, patience, tact, and rigid military justice than in small wars, and nowhere is more of the*

humane and sympathetic side of a military force demand-
ed than in this type of operation."[143]

In addition to the situations described in the manual, there are many other occasions in which Marine NCOs are required to exercise judgment where lives are not so immediately or obviously at stake. That doesn't mean the judgments made in those situations are any less pivotal to the Marine involved. A bad call made with faulty information or snap judgments based on emotional response can be the difference between saving a good Marine and creating one with an infectious bad attitude. Corporal Jennilee Leston has seen the value of judgment on a number of occasions where she had to call the shot and live with its impact. She also volunteered for service with a Female Engagement Team in Afghanistan, but her education in judgment required by Marine NCOs began a couple of years before that.

Leston is a relatively new NCO from a small town in Ohio who earned her stripes after she graduated from Correctional Specialist training and was assigned to the base brig at Camp Pendleton to guard Marines and sailors who ran afoul of regulations in situations serious enough to get them a stretch in correctional custody. She was assigned additional duties as the NCO in charge of annual training for all the other Marines in her unit. Marines are held to certain physical standards, such as weight and their appearance in uniform. One Marine in her unit had been on the Body Composition Program (BCP), a weight and body fat reduction program for Marines who fail to make minimum requirements. Their careers in many ways—including promotion or potential discharge—depended on judgments Corporal Leston had to make about motivation and dedication. It was a big responsibility for a young NCO.

One morning at Camp Pendleton one of her Marines arrived late for required formation with the excuse that he hadn't heard his alarm. Leston understood the situation and knew it was not an uncommon thing. She also understood that she had to judge whether it was a lapse or an excuse. She wasn't interested in excuses and she wanted to be sure this particular Marine had both the stamina and reliability to respond in an emergency. She judged that the best way to find out was to put him under pressure and so she assigned the tardy Marine to a particularly arduous physical training program to begin that morning. The result helped her determine that the Marine in question had the right stuff to continue and to improve. Leston didn't know it at the time, but that same sort of information gathering, assessment and decision-making was on the horizon for her during a deployment to the Middle East, an unusual assignment for NCOs from her type of unit.

She volunteered for overseas service with one of the new Female Engagement Teams and was accepted into the program. She joined a FET working mostly in support of 3rd Battalion, 7th Marines operating in Afghanistan's Helmand Province. Leston and the other female Marines on her team had to understand the unique situation among Afghan women. In fact, they had to keep from making snap judgments based on knee-jerk emotional reactions to how differently females were treated and it wasn't easy.

There were few guidelines beyond what she'd read or been told by others who served in that part of the world.

Brigadier General Lawrence D. Nicholson, who served as the Commanding General of the 2nd Marine Expeditionary Brigade, Task Force *Leatherneck*, while Leston was deployed to southern Afghanistan had an incisive perspective on culture shock among his Marines. In an interview with the Marine Corps History Division after his return to the U.S., Nicholson recalled what he told his people preparing for a year in Afghanistan:

> *"...I could send you to ten Mojave Vipers [a predeployment program at Marine Corps Air Ground Combat Center 29 Palms in California], I could give every one of you a Ph.D. in the cultural anthropology of Afghanistan, and you're still not going to be ready for some of the scenarios that will unfold."*[144]

Leston and all the other Marines headed for Afghanistan that year knew that the judgments they would have to make—with confusing information in unfamiliar situations among the people of a foreign culture—would be both difficult and crucial. General Nicholson summed up that challenge in another part of his post-deployment interview:

> *"We had guys making fundamentally strategic decisions every day by their actions, by their conduct, by the manner in which they carried out their mission, by the way they interacted with the Afghan Security Forces....principles are principles for a reason. It's because they are transferrable, because they fit more than one scenario and how you treat the people, how you work with the people, how you earn their trust. You can surge troops and*

equipment, but you can't surge trust, confidence and personal relations. That has to be built up over a period of years."[145]

Leston and the members of her FET were working on shaky ground but they at least had the experience of many fellow NCOs who served in Iraq to rely on for inspiration. There were as many differences as similarities in the Iraq experience, but each experience counted on the young NCOs who had to make crucial decisions and exercise good judgment on the ground. Lieutenant Colonel Willard A. Buhl, who commanded 3rd Battalion, 1st Marines, served with and observed Marine NCOs in Iraq, leading Marines in a counterinsurgency environment with both high-intensity urban combat and humanitarian reconstruction. Writing in the Marine Corps Gazette, Buhl cited the impressive judgment of the NCOs:

> *"These sergeants and corporals do not have all of the educational advantages…but have consistently demonstrated that with the right leadership and training they are more than equipped to handle the complexities of the operating environment in Iraq. These highly perceptive and intuitive NCOs quickly and effectively bridged seemingly vast cultural differences, patiently and expertly training Iraqi soldiers and policemen while demonstrating the altruistic values we as Americans hold dear."*[146]

Those lessons from Iraq were transferred by veterans of that conflict to young Marine NCOs like Sergeant Fonseca and Corporal Leston, but the business of sound judgment has hundreds of other admirable examples woven into Marine Corps culture from other NCOs in earlier conflicts.

And those lessons resonate when they come from veterans such as Freddie Joe Farnsworth, an infantryman who served in Panama and in Operations *Desert Shield* and *Desert Storm* as an NCO with units of the 7th Marines.

Farnsworth was an aimless youth in Wyoming, mostly interested in partying and sports, when he attended a career day at his high school in Pinedale back in 1985. "All of a sudden, a man in an awesome-looking uniform walked in and stepped up to the podium," he recalled. "As I looked around, all the girls were giddy and I knew instantly that I wanted to be a Marine." He was just seventeen years old at the time, so enlistment would require parental approval. Farnsworth brought a recruiter home to meet his folks and that sealed the deal. When he told his dad that he wanted to be a Marine, his father looked him in the eye for the first time. Farnsworth could see that his dad was proud of the decision and that his son was showing the first sign of a more mature judgment.

As an NCO with one enlistment under his belt and sergeant's chevrons on his collar, Farnsworth found himself leading a squad over the berm between Kuwait and Iraq in 1991. They had just endured a short, sharp firefight with Iraqi defenders and had taken several prisoners when Farnsworth's judgment was challenged. One of the prisoners had booby-trapped himself, and when the explosive detonated, the prisoner also wounded two Marines in Farnsworth's squad. Things were getting ugly and there was more than a little sentiment for quick revenge against the remaining Iraqi POWs. Farnsworth had to curtail those emotions, which were running high amongst his men. He had to follow orders to form the prisoners into some semblance of order, strip them of any military gear, search for any concealed weapons or explosives and get them ready for processing to a POW holding area. It was clearly not going to be an easy thing to do with angry Marines on one hand and resentful, uncooperative Iraqis on the other.

"We were in the middle of getting them into formation," Farnsworth remembered of that tension-filled experience:

> *My squad had taken the bulk of the prisoners. We were in charge of about fifty of them, but in the middle of me getting the prisoners in formation, there was one that was very reluctant to follow any of my orders, and my security team was getting very edgy.*[147]

The balky prisoner finally got into formation but he was having no part of removing his clothes when the Marines, using a combination of what little Arabic they knew and some creative hand signals, demanded that the prisoners take off their outer clothing and gear for inspection. Other prisoners were watching closely how the Marines responded to this defiance and Farnsworth knew he had to make a decision in a hurry and it had better be the right one or blood might be spilled.

"My Marines were getting really uptight, but I knew that if I didn't keep control of the situation, it could result in a prisoner being harmed or even killed." The Marines were shot through with adrenaline from the recent firefight and Farnsworth knew their judgment might be hampered or adversely influenced if he didn't defuse the situation. He approached the defiant prisoner and made an impassioned speech that the man very likely didn't understand but he got the idea through Farnsworth's passion and creative pantomime. The man finally nodded and began to strip off his uniform.

"When I saw that, I thought that it was over," Farnsworth remembered. "Then I made a poor judgment call and turned my back to him for just a second. That's all it takes." The prisoner reached out to grab Farnsworth from behind. Feeling something, he turned quickly and in a flash, the prisoner was on the ground and a Lance Corporal had the muzzle of his M249 light machine gun up against the prisoner's nose. Farnsworth knew he had to maintain control of the situation immediately.

"I kept my cool and the situation calmed down immediately. The prisoner changed his attitude and complied with all our orders from that point on." Later investigation revealed that the man was a colonel in the Iraqi Republican Guard who had failed to convince his troops to fight to the death in defense of the position Farnworth's Marines had just taken.

Farnsworth learned a thing or two from that situation:

> *"The lesson I learned was that by keeping myself calm, it kept my troops calm as well. Our forces gained good intelligence from the prisoner, who could have easily been killed that day. It's an example of how one poor judgment could have changed the outcome for a lot of people."[148]*

Leaders must use their best judgment to maintain control of themselves and their environment. Their subordinates are watching to see how the leader judges the situation, so that they know how to react. Had Farnsworth not kept calm, his Marines could have interpreted his anxiety as permission to fire, which would have been a tragic conclusion to his story.

Lessons in judgment and its consequences like the one learned by Sergeant Farnsworth that day in Iraq have been passed along through the Corps from NCO to NCO, sometimes formally and often word-of-mouth from one generation of small unit leader to another. It is all part of a continuing study of leadership that begins from the time a Marine graduates from recruit training until he departs the ranks as Farnsworth did as a Staff Sergeant a few years after his experiences in *Desert Storm*.

A generation after Farnsworth and his Marines fought through the desert sand and flaming oil wells during *Desert Storm*, Sergeant Kent Hedgepeth got a practical application course in battlefield leadership and good judgment in Iraq. He served in Operation *Iraqi Freedom* through major battles in places like Ramadi and Fallujah where he served as an NCO with 3rd Battalion, 2nd Marines. His judgment and that of his fellow leaders was challenged every day during that long, brutal deployment as they sought to defeat a resilient, capable enemy and in trying to build trust among the Iraqi people. It was a constant challenge to strike a balance between judgments made in combat circumstances and those made in trying to build effective relations with Iraqis brutalized by years of war in their backyards.

"You can't forcibly take what we're trying to gain: trust and confidence," Hedgepeth said. [149] To build confidence leaders need understanding that will lead to sound judgments. That implies an understanding of the nature of judgment which is a difficult proposition in and of itself.

Having knowledge allows leaders to understand what options are available when opportunities and crises arise. But just knowing those options is not enough; the leader must be able to evaluate which of those options will, when enacted, produce the best results.

Judgment is a contextually-informed process that leads to making decisions. It involves sorting through all available knowledge, sometimes quite quickly, and gleaning out what is important, what is trivial, and what is useless, even if it is interesting and distracting. One roadblock to judgment is an overabundance of information, which can complicate the process of discerning the most important pieces, especially if some of them are of personal or emotional interest. This decision paralysis happens when too many options, even good ones, freeze the ability to make judgments. When overloaded with options, "choice no longer liberates, but debilitates. It might even be said to tyrannize."[150] Judgment is the tool that evaluates options and eliminates enough options that are likely to produce poor outcomes that the process of deciding between them becomes manageable.

Judgment also is used to determine whether or not a decision in fact needs to be made. Some decisions are vital while others can be delayed or avoided all together. Many leaders have become distracted into making a variety of relatively unimportant decisions while letting the tough ones languish, feeling as though they are making progress when in fact, they are stagnating. Other times, a judgment of an action may be correct, but the timing may be off.

One critical component of judgment is timeliness. Often, leaders must assess situations quickly and without significant time to reflect. The Marine Corps refers to the "70 Percent Solution," meaning an imperfect solution that can be acted upon quickly, rather than waiting for the perfect judgment—which may never come. This guideline doesn't advise acting in extreme haste; rather, it advises avoiding "analysis paralysis." It argues that with 70 percent of the possible knowledge, having completed 70 percent of the analysis, and with a confidence rate of about 70, then the time is right to make an informed judgment.[151]

By choosing not a perfect judgment, but the best judgment that can effectively address the issue at hand in a timely manner, it also leaves room to adapt and adjust as new knowledge comes to light.

Chapter 8

Enthusiasm

*It is not the critic who counts; not the man who points out
how the strong man stumbles, or where the doer of deeds could
have done them better. The credit belongs to the man who is
actually in the arena, whose face is marred by dust and sweat
and blood.... but who does actually strive to do the deed; who
knows the great enthusiasm, the great devotion, who spends
himself in a worthy cause, who at the best knows in the end the
triumph of high achievement and who at the worst, if he fails,
at least he fails while daring greatly. So that his place shall
never be with those cold and timid souls who know neither
victory nor defeat.*
–Theodore Roosevelt (1858–1919),
Man in the Arena Speech given April 23, 1910

What is enthusiasm in a leader? As with most other
leadership traits, there's no single, one-size-fits-all
answer but Captain Baron A. Harrison had some thoughts
on the matter in an article he wrote for the *Marine Corps
Gazette* in which he described a company commander he
particularly admired. "It was plain to see that he loved
being a Marine. No situation would ever get him down. He
never displayed any doubt in himself or his company."
Harrison used as an example a cold, rainy night after his
unit had conducted a number of helicopter-borne raids in a
particularly arduous training exercise. Everyone was wet,

chilled and miserable when they discovered they'd have to conduct yet another foray into the dark. Enthusiasm was rock-bottom when the company commander assembled the junior officers to brief the mission.

"What is the matter with you guys? Aren't you excited about this raid?" It was clear that no one was but that didn't dampen the CO's attitude. He conducted an upbeat, enthusiastic briefing that had them all grinning and some even anxious to get on with the ordeal. "He had not said anything particularly inspirational," Harrison wrote, "but his enthusiasm was clear from his mannerisms and tone of voice." The officers left the meeting laughing, motivated and ready to roll. This renewed enthusiasm was transmitted to their NCOs who used it to buoy the spirits of the Marines who would have to conduct yet one more miserable exercise on a miserable night. Harrison noted that, "In the end, our CO's energy turned what would have potentially been a 'go through the motions' type of evolution into a meaningful one that we all felt good about conducting."[152] Enthusiasm is infectious and it spreads by contact.

The origin of the word "enthusiasm" springs from *en thusia*, the Greek word for "in sacrifice." This origin often has been interpreted as "religion," because sacrifice was an important part of ancient Greek religions[153] and because of the Greek word for God: *theos*. But the significance for leaders comes closer to the original interpretation.

When we're enthusiastic about something, we're willing to sacrifice for it. People who are enthusiastic about a cause will sacrifice time and money for it. People who are enthusiastic about their jobs will make personal sacrifices to spend time at work and educate themselves to do a better job. Men and women who are enthusiastic about being Marines understand that sacrifice might come at a very high price.

Rafael Peralta paid that price. As a sergeant with A Company, 1st Battalion, 3rd Marine Regiment, Peralta served as an NCO during Operation *Dawn*, an offensive to recapture Fallujah in 2004. Those who knew him best recalled him as the sort of Marine who loved what he was doing and as a leader who clearly loved the Marines he led. His background explains a lot of that attitude.

Peralta came from a Mexican immigrant family living in San Diego. He earned citizenship after joining the Marine Corps. Peralta served the United States with enthusiasm. In his parents' house, only three items hung on his bedroom wall: A copy of the United States Constitution, a copy of the Bill of Rights, and a copy of his Boot Camp Graduate Certificate.

When he became an NCO, Peralta rapidly gained a reputation as a man who frequently put his safety, reputation and career on the line for the needs and morale of the junior Marines around him.[154] That was demonstrated nobly on November 15, 2004 when Peralta and his Marines were ordered into Fallujah to help clear the city of insurgents during a brutal, bloody street-fight that involved clearing buildings block-by-block while under heavy fire from fanatical defenders.

As they entered yet another house deep in the city, Peralta's Marines kicked their way through the doors to two rooms that turned out to be empty. Peralta personally forced his way into a third room and was met by a hail of full automatic fire that ripped into his face and upper body. He spun aside to give his comrades a clear shot at the three insurgents in the room who wounded him. Those defenders responded by pitching out a fragmentation grenade that landed near Peralta and the other Marines. That grenade would have killed many of them but Peralta was having none of that. Although barely alive, he reached for the grenade and pulled it under his bleeding body to shield his Marines from the blast.

Shrapnel from the detonation wounded one Marine seriously but none of the other Marines suffered more than minor scratches. One of those Marines from Peralta's squad, Corporal Brannon Dyer, told a reporter from the *Army Times*, "He saved half my fire team."[155] Lance Corporal T. J. Kaemmerer, a combat correspondent attached to the unit, remembers a friend saying: "You're still here; don't forget that. Tell your kids, your grandkids, what Sergeant Peralta did for you and the other Marines today."

Before he deployed to Iraq, Peralta wrote to his brother Ricardo, who then was fourteen: "Be proud of me....and be proud of being an American."[156] Peralta was clearly and demonstrably enthusiastic about serving as a Marine and he took every opportunity to let his family in and out of uniform know that. It was that infectious sense of doing something important as much as their training and loyalty that kept Peralta's Marines charging through the bloody streets of Fallujah. He was a hard man to ignore and an easy man to follow. That's a good working definition of enthusiasm for Marine NCOs.

Enthusiasm as an effective leadership trait is closely linked to judgment. A leader with enthusiasm is able to identify what is necessary, productive and exciting, and then embarks on the required work not with a plod but with a rush. He works through the action at quick, high-spirited pace and that sense of enthusiasm tends to lift the spirits of those laboring beside him.

Even when the requirements are difficult, if it is necessary or required in a given situation, enthusiastic leaders set aside the negative aspects and focus on the positive energy they can bring to the table. It's not easy. It takes more than a little self-discipline. But it works and a show of enthusiasm often leads to truly inspirational behavior.

But there's more to enthusiasm than simply good will and positive attitude, although that is a good start. Not everyone is going to respond to a simple sense of energy and urgency on the part of leader who is asking his team to undertake a difficult or demanding task. Part of enthusiasm is communication. If a selected task force doesn't understand what's required or the importance of something that's being asked of them, it can be difficult to work up much enthusiasm no matter how happy their leader appears to be with the mission. Strong leaders ensure their teams have enough knowledge and information to be able to see why the situation is exciting. And since people get more enthusiastic about things they have a stake in and things they can personally influence, great leaders involve as many of their subordinates as possible.

In some cases, the leader can provide some background on the situation that illustrates the importance of the unpleasant task and use his positive momentum to communicate a spirit that his is an outfit that gets the worst jobs done in the best manner. That spirit generates a sense of perverse pride and gives a unit the sense that they are the go-to team for the really tough tasks. In many cases this approach works wonders for building enthusiasm.

This kind of approach has withstood the test of time. In fact, Sergeant Louis Cukela was of those Marines who was able to master it. Cukela was an enthusiastic Marine NCO who fought in World War I and remained in the Corps following the armistice. Over the years, he became something of a legend. Cukela claimed to be able to speak six languages. What is less sure is how well he mastered any of them. His first real recognition as a leader came when he mixed a metaphor and told one of his men that he was more than a little displeased with the man's performance on a given task. Sergeant Cukela: "Next time I send damn fool, I go myself."[157]

Then-Sergeant Cukela was awarded the Army and Navy Medals of Honor for "extraordinary heroism" in combat on 18 July 1918 during the World War I Battle of Soissons, photograph taken between 1921 and 1930. (U.S. Naval Historical Center Photograph—Photo #: NH 79333. Public domain.)

Those words caught the imagination of practically everyone in the American Expeditionary Force in France. They made the rounds through a series of cartoons that appeared in military newspapers and other publications. The phrase ended up inspiring a series of cartoons, and was endlessly repeated up and down the chain of command.[158] It became such a catch-phrase that it has since been attributed to such show-business luminaries as Samuel Goldwyn and Hungarian director Michael Curtiz.[159] They may have said it, but likely because they heard it before in Cukela's classic comment somewhere along the line in their careers.

Next time I send damn fool I go myself. What's so enthusiastic about those words? Imagine Sergeant Cukela—or yourself for that matter—saying them with a smile and then leading a unit out to get the job done with a renewed sense of humor and sense of purpose. Cukela's fractured command of English grammar aside, he was simply showing enthusiasm for a task at hand and inadvertently being self-effacing in the process. It's hard not to appreciate that as evidenced by the popularity of the phrase

Louis Cukela, an ethnic Croat, was born Vjekoslav Lujo Čukela on May 1, 1888 in the Dalmatian city of Split, Croatia (at the time the Kingdom of Dalmatia, Austro-Hungarian Empire). He is one of a proud line of immigrants who have served the United States during wartime, many without benefit of citizenship. Cukela arrived in the United States in 1913, settling with his brother in Minneapolis, Minnesota. After a short tour as a corporal in the U.S. Army he was discharged and then enlisted in the Marine Corps after the outbreak of World War I.

Although technically an enemy alien, Cukela brought a healthy dose of enthusiasm to the table when he joined the Corps telling anyone who would listen that he wanted to serve his adopted nation and to topple the oppressors of his native people. He told his recruiting sergeant that his father was in prison in Croatia, serving hard time for expressing anti-German, anti-Austrian views. He added that other members of his family were already in the battle and that he wanted immediate service overseas. "I want to go and fight those Huns," Cukela said, thinking that the Marines would give him his best chance to get into the fight.[160]

He got his wish. The Corps sent Cukela to France with the 5th Marines and he took part in all the battles that regiment fought with an infectious enthusiasm. Cukela didn't just talk about his enthusiasm for the job at hand on the bloody battlefields of Europe in 1917–18. He proved it through actions like the one that earned him both the Army and Navy Medals of Honor for the same action near Villers-Cotterêts, France. His citation for heroism in that action paints a picture of a leader who battled through hard times and inspired the men who followed him to victory.

On the morning of July 18, 1918, during the Soissons engagement, the 66th Co., 5th Marines was advancing through the Forêt de Retz when it was held up by an enemy strong point. Despite the warnings of his men, Sergeant Cukela crawled out from the flank and advanced alone toward the German lines. Getting beyond the strong point despite heavy fire, Cukela captured a gun by bayoneting its crew. Picking up their hand grenades, he then demolished the remaining portion of the strong point from the shelter of a nearby gunpit. He took four prisoners and captured two undamaged machine guns. Cukela was wounded in action twice, but because there is no record of either wound at the Navy's Bureau of Medicine and Surgery, he was never awarded the Purple Heart.[161] Cukela's enthusiasm pushed him to heroic action that day. More importantly, it inspired a group of pinned down, dispirited Marines to follow him and win a fight that was crucial to the allied effort in that part of France.

Cukela never lost his enthusiasm or his understanding of what it meant to the Marines he led. In the 1930s, Cukela was attending the Army's Infantry School at Fort Benning, Georgia. Working on a practical application problem in infantry tactics, Cukela was once again a man with few but memorable words.

"I attack," he told the Army instructor. It was not the accepted solution to the problem being considered. On the contrary, the frowning instructor advised, the proper action was to withdraw to more defensible terrain and establish a hasty defense but Cukela was having none of that. "I am Cukela. I attack," he retorted and tapped the pale blue, star-studded ribbon on his chest. "How you think I get this?"[162] It was as difficult then as it is now to argue with that kind of enthusiasm.

In the late 1930s, Cukela took his enthusiastic attitude to Parris Island in South Carolina, where he ran the rifle range. On a memorable day under his watch, recruits on the firing line managed only pitiful scores in a string of rapid fire. Cukela grabbed the public address microphone and shouted: "Cease fire. Clear and lock your piece. Fix bayonets. Charge the butts!" Fifty shooters on the Parris Island firing line, confused but motivated by Cukela's enthusiasm, did just that. As the recruits were charging the targets with fixed bayonets, Cukela's voice chased them downrange: "You can't shoot them; you go stab them."[163]

Cukela's infectious enthusiasm for any assignment he was given and the effect it had on the Marines he led has become legendary. As late as 1999, Defense Secretary William S. Cohen made reference to Cukela's dedication and opined that it might seem unusual for a man who was born in another country but Cukela was a special case in many ways:

> *"What distinguished this valiant American is his heroism, not the fact that he was born in the Balkans. It was said that Sergeant Cukela "had little interest in his own ethnicity and during his life was called Austrian, Slav, Yugoslav, Serb, and Croatian." In fighting in a war that started in the Balkans, this young hero did not care to be caught up in the ancient grudges. He was fighting for democracy. He fought for his country. He fought for America."*[164]

Louis Cukela did that practically every day of a long career as a U.S. Marine during which he eventually became an officer. He received a field appointment to the rank of second lieutenant in the Marine Corps Reserve on September 26, 1918, and was selected for a commission in the regular Marine Corps on March 31, 1919. Promoted to first lieutenant on July 17, 1919, he was advanced to the rank of captain on September 15, 1921.[165] He retired as a major in 1940, but was recalled to active duty to serve during World War II. He wasn't overly-enthusiastic about leaving the Marines again, but the Corps finally retired him from active service in 1946.

Cukela was a mighty warrior who always looked out for his Marines, and who could be counted on to be there when a Marine needed a helping hand.[166] He was a unique and enthusiastic leader but his motivation in serving an adopted country is not all that unusual even today. In fact, about five percent of U.S. active duty personnel were born overseas.[167] Between September 2001 and May 2008, U.S. Citizenship and Immigration Services naturalized more than 37,250 foreign-born members of the armed forces, and granted posthumous citizenship to 111 of them who died in combat overseas. That's enthusiasm personified and many of these foreign-born service members maintain that drive and spirit throughout their service in uniform. According to USCIS Director Alejandro Mayorkas:

> *"Many of our service members have risked their lives across the globe before becoming citizens here at home. Their brave acts, and those of more than 65,000 service members who have become citizens since 2001, demonstrate an extraordinary commitment to America. We are enriched by their decision to serve our nation and to join us as United States citizens."[168] In addition, more than 20 percent of the Medal of Honor recipients in America's history have been immigrants to this nation.[169]*

Foreign-born Marines certainly are not the only Marines who display enduring enthusiasm, but they are an interesting group to examine for the motivation behind their decisions to serve. The promise of U.S. citizenship alone can be a compelling reason to enlist, but those who choose the Marine Corps, where combat deployments are a way of life, have made a choice fraught with particular risk. None of that dampens the enthusiasm of men like Erbol Bekmuratov, originally from Almaty, Kazakhstan. At age sixteen, Bekmuratov moved to Philadelphia, where his father had found a job.[170] He contacted a local recruiter as soon as he was old enough to enlist. "I chose the Marines because when I moved here, I heard they were the best this country had, and I wanted to belong to it."

Corporal Mervin Roxas is another "Green Card Marine" who chose the Corps after studying the situation when he moved to the United States from the Philippines with his family in 1996. He explained his enthusiasm for joining the Marines:

> *"I never thought about being a Marine while I was growing up. In 2001, I was in college, majoring in Criminal Justice, when the 9/11 attacks happened. I wanted to do something about it but I was not sure what to do. Then a Marine recruiter came to one of my classes and talked about his experience in the Corps. It was only then that I decided to do something for my newly adopted country which gave me and my family so much opportunity. I joined the Marines in March 2002. Although I love the Philippines with all my heart, the U.S. is my home now and I'm proud to fight for what America stands for. I chose to be in the infantry and I don't regret a single day."[171]*

Here's the part Roxas glossed over. His first deployment to Iraq in 2003 was fairly uneventful. But when he returned to that combat zone in 2004, his unit had the tough task of training Iraqi military formations and running patrols near the Syrian border. That deployment got hot and heavy in a hurry. On July 5, they were hit hard.

"All I remember is that we were hit and I blacked out for maybe a couple of minutes," Roxas told the *Los Angeles Times*. Thinking they were still under attack, Roxas picked himself up, found his machinegun, and prepared to get back in the fight. "I didn't realize I'd lost my arm," he said.[172]

Along with his arm, Roxas lost part of his shoulder. His jaw was shattered. He was sent to a hospital in Germany although he has no memory of that time. He then spent more than a year at Walter Reed Army Medical Center in Washington, D.C., undergoing multiple surgeries. Although he had been studying criminal justice, there is not much future in law enforcement for a twenty-one year-old amputee.

"Crossing your arms. I really miss that," Roxas said. Losing an arm as well as a long-desired career can put a big dent in enthusiasm, but Roxas has lost none of his. He said he has no regrets about serving and would do so again if he could: "A lot of good things have happened. I have a better perspective on life, and I wasn't the most mature person when this happened."[173] And it all happened before Roxas was granted American citizenship: "I was very lucky to serve with great Marines that showed us how to lead men in combat." Even after sacrificing his arm, Roxas retains his enthusiasm for the Marine Corps, the Marines he led, and his new country.

Enthusiasm isn't all smiles and cheerleading. When an NCO cheers on an initiative that is sub-par, he may believe he's being enthusiastic, and that it is his job to push new policies so that his Marines are excited about the additional duties and keep up their motivation. Sometimes, however, initiatives are not particularly well considered. Sometimes what looks good on paper hits snags in the field. A good NCO is able to discover those problems and communicate them so adjustments can be made. While he can't and shouldn't criticize in public, the solid leader can use his enthusiasm to take appropriate action and let seniors know about problems so adjustments can be made. That's a display of enthusiasm for the big picture that helps eliminate problems down in the ranks where the little picture counts most.

Marines serving under a falsely enthusiastic NCO generally are going to be sharp enough to recognize when an initiative is not excellent, regardless of an NCO's cheering. That leader will then find himself marginalized as his unit figures that their leader either didn't have the judgment to know what a good initiative was or that he was just putting on an act for them. This kind of false enthusiasm makes leaders appear superficial and disingenuous. When that happens, team members feel a lack of trust, believing that their leader either didn't trust them to handle the truth, he didn't trust that they were smart enough to be able to judge the policy on their own. And since a real dialog about what was wrong with it or how to improve it never took place, the initiative was not improved. In the end, the NCO's reputation was harmed from above and below.

Honest enthusiasm understandably comes from an honest approach to a task or a mission. It's not only infectious; it's also something that builds trust and confidence between leaders and those they lead.

Corporal Robert Garcia is a Marine NCO who understands honest enthusiasm, and he displays it in and out of uniform. He was born in Mexico and moved with his family to Southern California, where he attended high school. He enlisted in the Marine Corps while still in school and shipped out to recruit training after graduation. Garcia was deployed to Iraq with a unit of the 1st Marine Division in 2003 for early combat service in Operation *Iraqi Freedom*. He was anxious to go and enthusiastic about being a leader. It was the fulfillment of what he considered a sacred obligation to his adopted country and he was not shy about letting his fellow Marines know about it during the deployment where he was wounded in action during the lightning drive on Baghdad.

"By birth I was born Mexican," Garcia said, "but I'm glad I got to be a part of the extended family: the Marine Corps. I get choked up over this." When Garcia was younger, his mentors were all Marines, and that's the only thing he wanted to be. "If you actually want to make a difference, if you want to feel like you're a part of something, the Marine Corps is going to be the way to go."

That wasn't an original idea in the crowd Garcia ran with before he enlisted. An enthusiastic friend joined a year earlier and gave him a glowing report. "It's not about money; it's not about uniforms," Garcia's friend said when he came home on leave. "There's just something different. The Marine Corps is going to offer you something that no one else is going to be able to offer you." That was good enough for Robert Garcia, and he never lost similar enthusiasm, even when his unit was headed for combat to drive invading Iraqi forces out of Kuwait:

"When we were heading into combat, I would say: 'I'm ready to die; let's do this.' And the others would tell me I

*was crazy. But I say you cannot win if the other side has
more will than you do. If they are willing to die and
you're not, you'll lose. You have to have the will to match
theirs. This attitude doesn't mean I want to die. But I
have to be willing or we'll lose. If your enthusiasm to do
your job does not match up with theirs, you're going to
lose. Point blank."[174]*

Garcia was granted American citizenship after his return
from Iraq. He recalled that the judge who administered the
oath was nearly as enthusiastic as the new citizens who
were raising their hands to swear allegiance to a new
country. "I love Marines," the judge said. "People forget all
of what you do. We need more citizens like you—you're
paying it forward before it's even paid to you."

Garcia took a discharge from the Corps after his enlistment was up, but he still has enthusiasm for the Marine
Corps and the people who remain in the ranks:

*"My loyalty is first to the Marine Corps just because they
gave me the opportunity to take care of my family. Everything that the Marine Corps gave to me deserves my biggest loyalty in return."[175]*

Gunnery Sergeant Rafael Jimenez, the Recruit Training
Regiment equal opportunity representative, said those who
serve honorably deserve the best gift the country can give,
which is to grant them their citizenship:

*"Although foreign recruits grew up in different places
than others in their platoons, cultural differences do not
tend to interfere with the routine of training. In fact,
immigrants' unique backgrounds often help their success
through the rigors of Parris Island."[176]*

Sergeant Jorge Silva, another modern-day Green Card Marine, said that the leaders he served under all had infectious enthusiasm:

> *"I had a Gunny, who I'm sure picked up part of it from a quote or a movie. He would say that 'Not only is every day a holiday, every meal a feast but every paycheck is a fortune and every formation a parade.' All the Marines I served with not only would show a high level of enthusiasm but they could put a bit of humor to a situation to motivate the rest of us, no matter if it was rain or shine outside. We could have had three hours of sleep but our sergeants could inspire us to jump into action. What I always remember about my leaders were endurance, enthusiasm, and inspiration."[177]*

Sergeant Jose de la Cruz remembers his father trying to enlist back in the 1970s. He had moved to the United States from Mexico, acquired his green card, and was able to live the life he wanted. He wanted to give something back, to pay for his citizenship. He tried to join every branch of service, but once they learned he had three kids, each told him that he was disqualified from service. His father's inability to serve always stuck with Cruz, and part of the reason he joined the Corps was in gratitude for the life that he and his family were able to have. Throughout his service and beyond, de la Cruz kept his enthusiasm, which he believes is a winning characteristic of Marine Corps NCOs:

> *"I think that if the NCO truly believes in what he is doing, he will be enthusiastic about his duties. Always accepting the challenge to do the most uncommon*

and unnatural thing and being optimistic and always looking for the next challenge: that is what makes the Marine NCO superior.

"To be enthusiastic about your life as an NCO, we accept leadership the Marine Corps way, which involves all fourteen leadership traits. If one trait is missing, that perfect circular sphere stops revolving, causing it to wobble. I look at Chesty Puller [one of the most iconic of Marine Corps generals, who started his career as an enlisted man] in the 'Frozen Chosin,' in the Korean War. He accepted the challenge, was optimistic in my opinion, when he said: 'All right, they're on our left, they're on our right, they're in front of us, they're behind us ... they can't get away this time.'"[178]

From Cukela to Roxas to de la Cruz, Marines have demonstrated not just mere commitment to their jobs, but enthusiasm for them as well. Enthusiasm has helped them exercise judgment in positive ways as they sought to make the right decisions under challenging circumstances. But enthusiasm alone doesn't decide leadership. How Marines communicate that enthusiasm tactfully also is crucial to successful leadership.

Chapter 9

Tact

A leader is the man who has the ability to get other people to do what they don't want to do, and like it.
—Harry S. Truman (1884–1972)

"Come on you sons of bitches—do you want to live forever?"

The immortal words of Gunnery Sergeant Daniel Daly urging his men forward into brutal German fire sweeping the bloody wheat fields before Belleau Wood during World War I have a rock-solid place in Marine Corps lore and legend. Historians argue whether that single, profane shout turned the tide at Belleau Wood and spurred a swarm of bayonet-wielding Leathernecks into an attack that carried the day back on June 18, 1918, but Marines don't question the effectiveness of Daly's leadership on that pivotal day in a crucial battle.

There is little question that Daly's words were effective, and the history books bear witness to that fact. Those Marines got up, charged, and did what was necessary. But was Daly, a leadership icon in the Corps, being tactful? Arguably, yes, but that might seem counter-intuitive given his choice of words—depending on how one defines "tact."

Tact is a strange leadership attribute coming from the U.S. Marine Corps: the in-your-face, profane, salty and aggressive stereotype of a Marine seems at great odds with the usual perception of tact. Isn't tact being polite, diplomatic, inoffensive, and politically correct? Not always. Tact may involve those things if a given situation calls for them, but often—particularly in military circles where plain speaking is considered a virtue—it doesn't. Tact is the ability to communicate in the language that best allows a listener to understand the message or meaning that's being communicated and to be motivated to act upon it. By that definition, Dan Daly was profoundly tactful that day nearly a century ago at Belleau Wood.

Tact allows communication, even between aggressive parties.[179] It facilitates comprehension of both facts and underlying emotions. People who truly desire to influence the thoughts or actions of others, especially in a compressed period of time or under other stressors, often can't avoid upsetting some feelings. That's mitigated by knowing and understanding people and their motivations. Given that background, the tactful leader chooses the language or behavior that will help the people in his audience to motivate themselves.[180] That can be direct—but still sensitive.

Playwright and master wordsmith Wilson Mizner had some insightful thoughts on the topic: "Tact is the language of strength," he said. "It is the ability to say something or make a point in such a way that not only is the other person not offended; they are totally receptive." Humans, however intellectual, are primarily emotional creatures. Tactful humans recognize that and use that understanding to choose the right things to say or do at the right time in a given situation. Moreover, tactful communication usually prompts the behavior they want—or at least greases sticky situations in which emphatic order-shouting just won't work.

Despite a contrary image, Marines like Dan Daly demonstrated it in abundance. A young recruit who joined the 6ᵗʰ Marines in France during World War I was shocked to discover that his new First Sergeant was the legendary Dan Daly. "My God! Do you mean he's real? I thought he was somebody the Marines made up—like Paul Bunyan!"[181] In many ways, the kind of tact displayed by Daly is what built the Marine Corps legacy, and it's a large part of what keeps that legacy alive among Marines.

Most people don't become legends in their own lifetimes, but Daly did, although it was hardly his plan. There's no evidence that he ever sought the limelight or tried to glorify himself at any point during his long career as a U.S. Marine. In fact, Daly is on the record as repeatedly saying, "I was only doing my job. I wanted to be a good sergeant of Marines."[182] He did his job well, and he did it with a surprising amount of tact for a man involved in a rough and often violent trade.

Marine legends like Daly were most often painted with broad brush-strokes as larger-than-life figures who swaggered with bravado, swore like foul-mouthed demons, and devoured enemies and under-achieving recruits alike. None of that fit Daly, who was for the most part a relatively soft-spoken, self-effacing man. Born in Glen Clove, Long Island, in New York on November 11, 1873, he was a small man, just 5 foot, 6 inches tall, 132 pounds, with keen gray eyes, a dry sense of humor, and a decisive mind.[183] Not much is known about his early personal life, but we do know that his exploits from the Boxer Rebellion in China during World War I became legend. According to historian Edward A. Dieckmann, Daly would have abhorred any sort of cheerleading about his battlefield heroics or effective leadership:

> *"The plain, blunt fact of the matter is that Dan Daly's story needs no embellishment of any sort. His career as one of America's outstanding Marines—his daily routine as a leader of men, his built-in kindness, his modesty and quiet courage, his thoughtfulness and utterly unselfish self-sacrifice—need no building up from any source."[184]*

There is no denying that Daly was rough and tumble when it came to action, but that doesn't mean he wasn't tactful as well. Lieutenant Colonel F. E. Evans, Adjutant of the 6th Marines, who served with Daly in France, credited him with a large measure of the unit's success:

> *"I came into close and practically daily contact with him. He and the Regimental Sergeant Major, in my opinion had more to do with the fighting spirit of the regiment than any other two officers or men. Daly's influence on new officers and men was remarkable. He enjoyed the respect, confidence and admiration of every man in the regiment. For loyalty, spirit, and absolute disregard of fear, he was almost unique in the entire brigade and his devotion to his officers and to the men of his company was demonstrated time after time."[185]*

The Regimental Adjutant wasn't the only Dan Daly fan. Many of his senior officers recognized his leadership qualities. These qualities allowed Daly to get into the hearts of his Marines and fill them with the will to act.

When Marines have the spirit to fight and to overcome fear and act, that courage rises up the ranks as well. General Allen H. Turnage, who was the commanding Officer of the machine gun battalion of the 5th Marine Brigade during World War I, echoed the prevailing sentiments about Daly:

> *"He was quiet, unassuming—with never a thought of publicity or fame. I can see him now upon receipt of each of his many medals—stowing it away and saying to himself: 'Just another job.' He was a man of excellent habits and conduct. His nearest approach to profanity was an occasional emphatic 'damn.' Of all the Marine NCO's whom I knew well during 35 years of active service, Dan Daly was most outstanding."[186]*

Daly brought a lot of motivation to his assignment with the Marine Brigade in France during World War I. His desire to be a leader of Marines began before the turn of the century in what would go down in the history books as the Spanish-American War. There was serious trouble brewing in Cuba—just ninety miles off America's southern shores—but Daly couldn't have predicted what would happen any more than could most of his fellow Americans, most of whom knew little or nothing of the situation in the Caribbean in the late 1800s.

During the last decade of the nineteenth century, Cubans were doing their best to stir up an effective revolt against Spain which had promised—but failed to deliver—their colonial charges independence over ten long years. To put down a building revolt and the attendant violent clashes between Cuban guerillas and Spanish troops, the Spanish crown sent a ham-fisted governor general named Valeriano

Weyler y Nicolau to the island. The island's leaders were bent on denying Cuban guerillas the support of a rebellious population. Shortly after he arrived in Havana, the new Spanish governor ordered everyone living in certain areas of eastern Cuba to leave their home villages and move to "re-concentration camps" under Spanish military control.

It was a very unpopular move. The overcrowded camps bred epidemic disease. No outside food or supplies were allowed into the camps, leading to the death of tens of thousands of Cuban civilians.[187] The press (most notably William Randolph Hearst and Joseph Pulitzer in the United States) covered these atrocities, so they were well known to the American people. U.S. President William McKinley responded early in 1898 by sending the U.S.S. *Maine* and a supporting squadron of warships to Havana Harbor as a show of American displeasure and as a rescue force in case American citizens needed to be evacuated from the turbulent situation on the ground in Cuba. That was the genesis of a controversial event that stoked fire in the bellies of men like Dan Daly and thousands of other volunteers when war with Spain was declared that same year.

On the hot evening of February 15, an off-duty Marine slung a hammock topside aboard the Maine to avoid the heat in his quarters below decks. An explosion hurled him up and slammed him onto the deck. A second blast followed seconds later,[188] sinking the *Maine* and much of her crew. For all intents and purposes—minus only the official declaration that followed shortly thereafter—the Spanish-American War had begun.

The Navy board that investigated the sinking reported that there was no negligence on the part of the ship's crew, and that the explosions came from an outside source. Although the Navy didn't actually accuse Spain or torpedoing or otherwise having a hand in the sinking of the U.S. warship, Spain was insulted by the wide-spread belief that they were to blame for the disaster. Spain declared war on the United States on April 23. It was a short, sharp, mostly-mismanaged conflict that plunged America into wartime frenzy—and launched a long, illustrious career for Dan Daly.

Hoping to get into the fight as quickly as possible, Daly enlisted in the Marine Corps on January 10, 1899. But he was too late. While he was still in training, the war collapsed.[189] Feeling he had missed something noble and important, Daly wasn't about to miss any future conflicts, so he reenlisted in the Corps and set his sights on getting ready for the next fight, whenever and wherever that might be. He made the Marine Corps both a passionate pursuit and a personal life choice. In fact, it doesn't appear from the scanty records still available about his early days in the Marine Corps that Daly had much of a personal life outside the ranks. He never married and once told a reporter, "I can't see how a single man could spend his time to better advantage than in the Marines. Life in the Corps isn't so bad after you get the hang of things."

In fact, Daly very quickly got the hang of things in the small, tight-knit Marine Corps of his day. He was consistently promoted through the NCO ranks, starting with corporal in 1906 and finally reaching sergeant major in 1920.[190] He was offered a commission but refused with a terse comment repeated to this day by proud young Marine corporals and sergeants: "Any officer can get by on his sergeants. To be a sergeant, you have to know your stuff. I'd rather be an outstanding sergeant than just another officer."

Daly's first combat experience came in China in early 1900, when members of a number of ultra-nationalist guerilla bands led by the Society of Righteous Harmonious Fists, which quickly became known as The Boxers. They waged a violent campaign bent on extermina-tion of all foreign-ers—especially foreign missionaries and Chinese Christians—in foreign commercial enclaves throughout China. Western diplomats were sufficiently alarmed to request that troops be sent in for protection.[191] A Marine guard was sent to Peking to guard the Legation Quarter, protected by the Tartar City Wall. Defensive positions along that wall were "the peg which holds the whole thing together," according to the legation

Daniel Daly being awarded the *Médaille militaire* from the French government for his actions during World War I. (Public domain.)

officer who commanded the defense force composed of military units from all the foreign nations represented. The American Marines found themselves in the thick of the fighting.[192]

Captain Newt Hall, Daly's commander in Peking, reluctantly assigned the young NCO to a vital one-man outpost on the Tartar Wall. "I cannot order you to remain here, Daly," the officer admonished. "The situation is very dangerous." It was a tactful way of saying Daly was not likely to survive a long, dark night alone on the wall if the Boxers attacked. Daly reportedly just shrugged and made an equally tactful reply: "See you in the morning, Captain." [193]

The feared Boxer attack wasn't long in coming, and, as predicted, it hit right where Sergeant Daly held his outpost. Using rifle fire, bayonet and rifle butt, he fought off attack after attack. Astonishingly, he held off a swarming enemy until morning when reinforcements arrived. That action on the Tartar Wall brought Daly his first Medal of Honor. It would not be his last. On the few occasions Daly could be induced to talk about that night in China, he was typically self-effacing.

"Those Legation ladies were wonderful," he said. "They ripped up all their ballroom dresses to sew up sandbags for us-all kinds of colors. I never saw such fancy sandbags. Some of 'em were even trimmed with lace!" When asked about his courage, Daly replied: "Oh hell. [Thirty-three] of us Marines and sailors got the Medal of Honor during that scrap. Why pick on me?"[194]

Even with a Medal of Honor proudly displayed on his uniform, all was not smooth sailing for Daly after his time

in China. He still had a bit to learn about tact and professional demeanor off the battlefield. In 1901, Daly spent
time in the brig at the Boston Navy Yard. About a month
after his arrival at the Marine Barracks, he was court-
martialed for being intoxicated while on post. Three weeks
after his release, he was court-martialed again, this time for
abusive language toward the sergeant of the guard along
with drunkenness.[195]

These two stretches—surviving on bread and water with
full rations delivered only every third day, along with a loss
of three month's pay—kick-started Daly's study of leadership couched in Marine Corps terms. By 1910, he had
become what official observers called a "strict disciplinarian
who demanded and got complete obedience from his men
... popular as well as respected ..."[196] He might not have
been able to articulate it at the time, but Daly had learned a
lot about tact. Other observers remembered him from those
post-brig days as a man who never touched a drink, smoked
tobacco, or let a cuss word slip from his lips.[197] Dan Daly
had come to understand that demanding one thing from
your Marines and doing another destroys trust and under-
mines effective leadership.

Reportedly, Daly also was a proponent of Marine-speak,
the salty language of sea soldiers peppered with naval
terminology and nautical references. When Lieutenant
Colonel J. M. Sellers was a newly-commissioned 2nd
Lieutenant in 1917, he remembered sessions with Daly that
were filled with salty conversation.

*"As I recall, Dan Daly was a ranking noncommissioned
officer of the machine gun company. Even that early, he
was widely known as one of the great old Marines. So it*

was quite natural for a number of us very green young lieutenants to gather around Dan in the cool of the evenings after we had drilled all day....He particularly enjoyed using nautical terms with which we had not been familiar before."[198]

Daly had clearly learned the power of words and the language used to convey their meanings. He knew there was a pride borne of speaking in special way and understood that Marine-specific lingo could be a very effective communication tool. He was a believer in properly applied jargon.

Jargon exists for reasons beyond effective short-hand. It develops in groups where there are no more suitable common words for things specific to that group. Typically, jargon involves terms not used by the general public. Although jargon can facilitate communication within the group, it also can be exclusionary to outsiders. This is a second handy asset that comes with jargon. It says to outsiders, "This is our culture; this is our language. If you don't get it, you don't belong here." With that come exclusivity and the pride that engenders. Marines have a huge storehouse of jargon and special-use words.

Among Marines, for example, there are two ways to indicate agreement: "yes" and "aye." There's a practical reason for it that goes beyond salty jargon. "Aye" is an acknowledgement that an order has been received, understood, and will be carried out. "Yes," on the other hand, is the answer to a question. If a Marine is asked by a corporal with unusual lack of perception, "Are you are a female?" the female Marine would properly respond, "Yes, Corporal." If she is told to go get ready for an inspection, she'd say, "Aye,

Corporal." Those are two different responses with distinctly different meanings among Marines, and the jargon involved is important.

For instance, Marines abhor using the word "gun" in reference to anything other than artillery or other weaponry in which the barrel does not contain rifling. One reason is accuracy; if you mean your rifle, say so. Don't use a generic term. Be precise so your audience knows precisely what you mean. A benefit of insisting on correct terminology is that it forces the leader to stop momentarily and think before speaking. Is it a gun or a weapon? A floor or a deck? A wall or a bulkhead? The momentary pause gives the mind a moment to make a decision, to chose the most tactful term and to use the word or phrase that communicates most effectively.

Of course, a quick reading of virtually any professional manual or house organ reveals the danger of jargon-drenched phrases that can either block effective communication or send a message that's not intended. The key to effective and tactful communication is understanding your audience, so that it can be determined if jargon will ease communication, or add confusion to the message.

Language changes over time. Words that were once offensive may become innocuous over time or vice-versa. "Jerry-rig," for instance, is a phrase occasionally heard today. A gadget is broken? No problem—just jerry-rig something to fix it. Few beyond etymologists realize the term entered the English language as a slam on Germans, who were referred to as "Jerrys," primarily by the British during World War II. During that conflict, German forces became notorious for scavenging parts or inventing them, to keep vehicles running and equipment functional. That led to the term jerry-rigged for a slap-dash, homegrown method of fixing things in the U.S. military when parts or replacements were not available. So, is it tactful to use terms like jerry-rigged or "gypped" (derived from a derogatory term for light-fingered Gypsies) as common jargon? It depends on the audience and your grasp of tact.

Bear in mind that tact is not the same as political correctness. In fact, the idea of political correctness as it infiltrated American English developed in response to some people's extreme lack of tact. Political correctness originally was an attempt to compel more conscientious, less possibly offensive word choice. That was admirable at some level, but politically correct language evolved into a vocabulary driven by fear of audience reaction. Tact, on the other hand, is based on respect for the audience. Tact is behaving in an interpersonally supportive way, with empathy and without threat.[199] What all this means is that language is an important aspect of tact.

So what about that vulgar phrase shouted by Dan Daly back in 1918? Was that the proper language at the proper time? There's little question in the minds of men like Floyd Gibbons, who was in France as a war correspondent for the Chicago Tribune:

> *"A small platoon line of Marines lay on their faces and bellies under the trees at the edge of a wheat field. Two hundred yards across that flat field the enemy was located in the trees. I peered into the trees but could see nothing, yet I knew that every leaf in the foliage screened scores of German machine guns that swept the field with lead. The*

bullets nipped the tops of the young wheat and ripped the bark from the trunks of the trees three feet from the ground on which the Marines lay. The minute for the Marine advance was approaching. An old gunnery sergeant commanded the platoon in the absence of the lieutenant, who had been shot and was out of the fight. The old sergeant was a Marine veteran. His cheeks were bronzed with the wind and sun of the seven seas. The service bar across his left breast showed that he had fought in the Philippines, in Santo Domingo, at the walls of Peking, and in the streets of Vera Cruz. I make no apologies for his language. Even if Hugo were not my precedent, I would make no apologies. To me his words were classic, if not sacred. As the minute for the advance arrived, he arose from the trees first and jumped out onto the exposed edge of that field that ran with lead, across which he and his men were to charge. Then he turned to give the chare order to the men of his platoon—his mates—the men he loved. He said: 'Come on you sons-of-bitches! Do you want to live forever?'[200]

And what about word choice? Was it the right choice on that bloody day in a difficult situation? Gibbons and a lot of Marines who were pinned down in that wheat field would agree it was. Daly, on the other hand, likely had his tongue lodged firmly in his cheek when he responded to a reporter who asked about the famous quote. "You know a non-com would never use hard language," Daly told the newsman. "I said, 'For goodness sake, you chaps, let us advance against the foe.'"[201]

Daly's quip to the reporter notwithstanding, he had an intuitive grasp of tact. In fact, Daly was famous well before the events at Belleau Wood for understating the importance of tact. A classic example was an incident in Haiti three years earlier when Daly was serving with another legendary Marine, then Major Smedley Darlington Butler. The story goes that Daly said only two things before he launched into the action that resulted in his first Medal of Honor.

They were conducting a horse-and-mule mounted patrol in the Haitian badlands searching for Caco rebels when they were forced to ford a fast-running mountain stream at night in a driving rain. The rebels ambushed the Marines, wounding several and causing panic among the pack animals, one of which was carrying a vital machinegun. Major Butler asked Daly about the gun as he rushed around trying to organize a counterattack. "It was lost in the river, sir," Daly replied.[202] While Butler was left to digest this unwelcome response, Daly took action.

He discarded most of his equipment, darkened his face, and began a treacherous trek through the advancing enemy to reach the point in the river where they'd lost the pack animal carrying the machinegun and ammunition. When he located the spot, Daly plunged into the river, swimming, diving, and scouring the muddy bottom until he found the dead pack mule with the weapon still roped onto its back. He had to dive several times to cut the gun loose and gather sufficient ammunition.[203] With Cacos firing at him all the while, he brought the machinegun, tripod, and ammo back to the Marine defensive positions. "I've set up the machine gun, sir," he informed Major Butler. And that was all he said before using the gun to devastate the attacking rebels and break the Marines out of a very bad situation. With very few words surrounding some admirable action, Daly set a classic leadership example that day in Haiti. And leading by example is just as challenging today as it was back then.

"Setting the example, it's not always easy," commented Gunnery Sergeant Michael Grassel at Camp Pendleton, California: "I hate dressing like a Lieutenant. I hate wearing a collared shirt and a belt buckle. But we have to set the example for our young Marines because they are the ones that will follow us."[204] Daly knew that back in 1915 and 1918 and he would recognize the truth in Grassel's take on leaders setting an appropriate example for their subordinates. If an NCO decides to hit the clubs wearing wildly inappropriate clothing, his young Marines will watch and take note—and emulate that behavior. Grassl explains:

> *Just like if I acted like a pig, all my Marines are going to see my room and they're going to think, Well, Gunny says this but he doesn't do it. So why do I need to listen to him? Because I won't do anything I don't expect my Marines to do, I get from them the rapport, that instant discipline, and obedience to my orders.*[205]

Grassl and NCOs like him obviously have an interest in their Marines and what they think. In many ways, that's an example of tact in action—even if it can be daunting to have men and women who are still teenagers under your care and supervision, especially during combat operations. But Grassel understands that is both a responsibility and an opportunity—a chance to set an example and lead by example in order to train younger Marines.

Leading by example requires tact, and tact requires harmony between what is said and how something is said. Words, tone, and body language all have to be in sync. When demeanor says one thing, but language says another, listeners will distrust what is said.

Authenticity is a key to tact. So is trust.

Building trust between the NCO and the Marines who serve under him is vital. Subordinates need to know that their leaders are going to make good decisions that are going to keep them alive whenever possible. They also trust that they are being prepared to step into their leaders' shoes if something happens to take that leader out of the equation. It's neither easy nor pleasant when the communication is negative in nature, when reprimands must be made, or when bad news must be delivered.

Leaders must tell the good, the bad, and the ugly and find the right words to avoid deliberately being offensive. Having to critique poor performance, inform them of an accusation against them, or discuss personal hygiene, for example, are all awkward situations. Most people naturally shy away from them, hoping the problem will go away or that someone else will step up and handle the situation. But problems ignored or negative situations avoided usually cause more trouble, and like Dan Daly in Haiti, good leaders will have none of it.

Another less contemplated aspect of tact is silence. Knowing when to shut up can be an important part of tact. Our brains are wired so that we can communicate in real time. It is not always necessary to filter your thoughts before you speak, although it's generally a good idea. If an immediate warning is needed, like yelling "Take cover!" in combat, naturally you need not spend any time pondering the right words or a more polite way to shout the warning. On the other hand, saying something spontaneously can easily lead to regret. People do this most often under stress, such as during a confrontation or an emotionally charged conversation. The trick is to know when to let thoughts become speech slowly and naturally, when to listen, when to speak, and when to keep silent.

Major David Anderson, an assistant professor at the U.S. Naval Academy, had this advice for young Marines:

> "… *listening does not mean simply maintaining a polite silence while you are rehearsing in your mind the speech you are going to make the next time you can grab a conversational opening. Nor does listening mean waiting alertly for the flaws in the other fellow's arguments so that later you can mow him down.*"[206]

Listening, speaking, and choosing the right words for the situation are the cornerstones of tact. Dan Daly is just one of the many Marines who have exercised strong judgment and demonstrated tact again and again, even under stressful situations. Being tactful comes with training and maturity, but it's also determined by making the right decisions—the right decisions about what to say, when to say it, how to say it, and who to say it to. Decisiveness is yet another of the characteristics that Marine Corps NCOs exemplify every day.

Chapter 10

Decisiveness

"In any moment of decision, the best thing you can do is the right thing, the next best thing is the wrong thing, and the worst thing you can do is nothing."
—Theodore Roosevelt (1858-1919)

With reports of modern-day pirates capturing vessels, stealing cargo, and even killing yachtsmen, it would be easy to believe that piracy is a problem unique to today. But, in fact, the situation facing America during the first part of the nineteenth century was considerably worse. The scourge of modern pirates in places like the southern Philippines and off the horn of Africa are nothing new, and back in the day of wooden ships and iron men, the American government, which was plagued by pirate murders, kidnapping, and thievery, took decisive action when nautical commerce was threatened.

That action most often involved U.S. Navy squadrons and detachments of Marines who were prepared to do battle at sea or on land against bloodthirsty pirate bands and the nations that sponsored their activities. One of those Marines was Solomon Wren, a twenty-four-year-old sergeant from Loudoun County, Virginia, who volunteered to go along on a punitive expedition against the infamous Barbary Pirates in 1804. His experience on that deployment became part of Marine Corps history and a classic example of the value of decisive action. A little historical background demonstrates the similarity between the situation then and what confronts modern Marines and sailors engaged in anti-piracy operations today.

Algiers, Tunis, Morocco, and Tripoli (known at the start of the nineteenth century as the Barbary States) had for generations demanded annual tribute from non-Muslim vessels that wished to trade in the Mediterranean. When America was a colony of Britain, American vessels were protected by the tribute paid by the British crown. But the new nation of the United States would have to pay tribute in its own right.[207] President John Adams reluctantly agreed to the expenditure, although the situation was fragile and the Barbary States continually increased their demands.

Adams and Thomas Jefferson, then minister to France, both attempted to negotiate with Tripoli's envoy in London, Ambassador Sidi Haji Abdrahaman. The envoy noted that in was written in the Qu'ran that:

"all nations which had not acknowledged the Prophet were sinners, whom it was the right and duty of the faithful to plunder and enslave; and that every Muslim who was slain in this warfare was sure to go to paradise."[208]

These words were written before colonialism had spread to Islamic lands, before oil interests influenced political decisions, and long before Israel had been founded.[209] Regardless, it created a situation that any perceptive modern citizen would recognize as familiar.

Great Britain encouraged the Barbary States, as the government there saw profit and advantage in motivating merchants to use British ships protected by the British Navy. In 1793, a treaty between Algiers and Portugal—arranged by the British—allowed Algiers to operate their warships in the Atlantic. This treaty led to the birth of the permanent U.S. Department of the Navy in 1798[210] in order to prevent further piracy attacks upon American shipping and to try to stop the onerous demands for tribute from the Barbary States.[211] Piracy was becoming a national crisis in America. The French were ambushing American trade vessels in the West Indies. And the tribute demanded by the Barbary Pirates had mushroomed to millions of dollars each year, including both the annual protection fee and ransom for Americans captured at sea, about a fifth of the entire national income.[212] In May 1801 Yusuf Karamanli, the Pasha of Tripoli, declared war on the United States, feeling the American tribute was no longer sufficient.

In 1803, once the United States had enough Naval strength, a squadron was sent to the Mediterranean, commanded by Captain Edward Preble. After a show of power, he left the frigate *Philadelphia* to blockade Tripoli. Unfortunately, the Philadelphia grounded on an uncharted reef. Gunboats from Tripoli were quick to take advantage of the situation. Although no one was wounded or killed, 308 Americans, including 43 Marines commanded by 1st Lieutenant William S. Osborne, were captured and held for 19 months.[213] Although damaged, the *Philadelphia* was kept anchored in the Tripoli harbor and used as a gun battery against the enemy. It could not be allowed to remain usable by enemy hands.

Navy Lieutenant Stephen Decatur, sixty-eight sailors, and eight Marines led by Marine Sergeant Solomon Wren were sent to destroy the Philadelphia. The task force entered Tripoli harbor on the night of February 16, 1804 aboard the USS *Intrepid*. Decatur knew the *Philadelphia* well; in fact, his father had once captained her.[214] The ship had been freed from the reef by sailors from Tripoli, and renamed the *Gift of Allah*.[215] It sailed into an anchorage under more than 115 cannons mounted along the walls of a castle that overlooked the harbor. The *Intrepid* also was moored at that anchorage, near twenty enemy vessels bearing cannon and armed crews. It was not going be easy to reach the *Philadelphia*, much less scuttle the vessel or set it on fire.

Fortunately, Decatur and his attack force had the advantage of subterfuge in their vessel. The *Intrepid* was originally a Tripolitan vessel that had been captured by Decatur and his crews in late 1803. She was a sailing under a different flag but still looked like many other ships in the region. Decatur hoped his ship would not be immediately identified as an American vessel when he used it as a launch pad for the raid on Tripoli Harbor. And that action couldn't come soon enough for Sergeant Wren and the Marines aboard Intrepid.

Conditions aboard the ship were miserable. The captured vessel was not large enough to accommodate the number of men aboard her. Wren had to constantly bolster the flagging spirits of his Marines, who slept on a shelf placed atop the water casks without even enough room to sit upright.[216] The only rations were biscuits and water. A brutal Mediterranean storm, which blew into the area as they were sailing into action, had to be endured before they could even reach their action stations. Despite such difficulties, Decatur led his sailors and Marines through it all and approached Tripoli Harbor without raising any immediate alarms.

On the night of February 16, 1804, Decatur entered the harbor and brought *Intrepid* smartly alongside the *Philadelphia*. As the enemy sounded the alarm, Decatur yelled "Board!" and leapt over the side to land on the decks of the target ship. Sergeant Wren led his Marines right behind their Captain, overwhelming the guards with sabers and tomahawks.[217] Combustibles were placed at key spots around the ship and ignited at Decatur's command. The fire spread rapidly and uncontrollably. As the enemy's gunboats and shore batteries began to fire, Wren led his Marines safely back aboard the *Intrepid*. Not one American was wounded. All but two of the enemy were killed or captured.[218] The raid lasted only twenty minutes, but it had far-reaching effect on nations which also were dealing with piracy in the area.

British Admiral Horatio Nelson called the raid to burn the *Philadelphia* "the most bold and daring act of the age." Pope Pius VII said the American action "had done more for the cause of Christianity than the most powerful nations of Christendom have done for ages."[219] Bold, decisive action in a key offensive operation had robbed the power of the Barbary Pirates, but it didn't end their activities.

The war between the United States and Tripoli continued and involved other bold strokes against the Pasha and his force of pirates struggling to control the seas in the area.

In one legendary feat of arms, Marine Lieutenant Presley O'Bannon led a mixed landing party including U.S. Marines, Greeks, Arabs, and Turks, on an arduous six hundred-mile trek across the desert to the harbor of Derna to attack the fort. O'Bannon and his Marines were the first American forces to raise the American flag in the Old World.[220] The Marines' Hymn commemorates these actions with the line " ... to the shores of Tripoli." And, in honor of O'Bannon's victory at the Battle of Derne, The Viceroy of the Ottoman Empire presented him with a Mameluke-style sword on Decemberr 8, 1805. [221] Reproductions of that sword are still carried to this day by Marine officers. Victory was all about decisive action in the war against pirates back then as it is today and Marines promote an appreciation for such bold strokes.

Decisive action is a long-standing tradition among Marines. *The Handbook for Marine NCOs* has some advice for modern Marines that would be familiar to predecessors like Sergeant Wren:

> *"Make sound and timely decisions. To make a sound decision, you should know your mission, what you are capable of doing to accomplish it, what means you have to accomplish it, and what possible impediments or obstacles exists (in combat, these would be enemy capabilities) that might stand in the way. Timeliness is almost as important as soundness. In many military situations, a timely, though inferior, decision is better than a long-delayed, though theoretically correct, decision."[222]*

General Krulak, former Commandant of the United States Marine Corps, said in 1999 that he intended to meet the challenges of the twenty-first century "by creating Marines and their leaders who have superb tactical judgment and are capable of rapid decision making under physical and emotional duress..."[223] It is an admirable goal, one that is easier to reach today than it was in the days of combat with the Barbary Pirates.

Information flows with amazing speed and in copious quantities on modern battlefields between individual leaders and operational centers that direct their activities and circle back to individuals in a fast, continuous loop. That presents challenges as well as advantages to modern Marines that would be unthinkable among their predecessors. General Krulak felt that: "Our leaders must be able to 'feel' the battlefield tempo, discern patterns among the chaos, and make decisions in seconds much like a Wall Street investment trader, but with life threatening consequences."[224]

That means the key to success is rapid decision-making. Decisions are made in several ways. The first method, which is what generally is stressed in formal education, is analytical decision making. Using this method, the decision maker gathers all the possible facts, weighs as many options as possible, and attempts to maximize the outcome. This is a perfectly legitimate method but it is time-consuming and cumbersome, and therefore not the best bet during the heat of battle.

Intuitive decisions are more common in those circumstances and are primarily based on the leader's judgment and knowledge. Research by psychologist Dr. Gary Klein indicates that most people make decisions intuitively rather than analytically more than 90 percent of the time[225] Despite such guidance, little is done to train people to make intuitive decisions well. General Krulak understood that and wanted the situation corrected in Marine Corps leadership instruction: "History has repeatedly demonstrated that battles have been lost more often by a leader's failure to make a decision than by his making a poor one."[226]

Realizing the truth of that assessment, Marines have adjusted their training and placed heavy emphasis on teaching leaders to make rapid, intuitive decisions, especially under pressure in training exercises where they are tired, cold, hungry and generally miserable. There's always an instructor at hand in these stressful exercises berating the leader in training with the constant question: "What are you going to do next? Decide now!"

Training designed to develop decisiveness has progressed into more sophisticated realms such as the Combat Decisionmaking Range, developed by the Marine Corps' Warfighting Laboratory at Quantico, Virginia, a branch of the Marine Corps Combat Development Command. The CDR's stated purpose is to improve current and future naval expeditionary warfare capabilities across the spectrum of conflict for current and future operating forces. The CDR is for use by every infantry regiment in the Marine Corps. Computer-based training went operational in March 1999, using seven major conflict or combat scenarios to train NCOs in combat decision making and leadership.[227] Krulak is a believer in the training:

"The CDR puts the squad leader square in the middle of the three block war and requires him to make decisions

across the spectrum of conflict, from humanitarian relief to mid-intesity firefights, with the media watching."[228]

According to Marines who have been through the training, the CDR improves their decision making skills and builds confidence in their intuitive abilities. They come away from the training much more capable and confident that they can make crucial decisions quickly. Although the technology may be new, however, decisiveness is hardly a new or unfamiliar pursuit in Marine Corps circles.

In 1989, the Marine Corps' maneuver warfare doctrine began to investigate U.S. Air Force Colonel John R Boyd's "OODA Loop" in attempts to discover how and why rapid decisions are made, and how they can be made even more quickly. OODA is an acronym for Observation-Orientation-Decision-Action, which describes the basic sequence of the decision making process.[229] Boyd, once a fighter pilot, began developing the concept by studying air combat during the Korean War. American aviators were quite successful compared to their enemy opposite numbers with a 10:1 kill ratio over North Korean and Chinese opponents.[230] Boyd wanted to understand why American pilots did so well and if whatever gave them the edge could be applied in other circumstances. The outcome of his study is what came to be known as the Boyd theory or the OODA Loop.

The OODA Loop applies to any two-sided confrontation, whether the antagonists are individuals or large groups. Each side begins the process of decision-making leading to a course of action by observing himself, the physical circumstances and surroundings in which he finds himself, and the opposing force. On the basis of that observation, he orients himself by making a cohesive mental picture of the circumstances. He makes estimates, assumptions, and judgments to determine what the situation means to him. Based on this orientation, he makes a decision and then puts that decision into motion by acting.[231] Since his actions have altered the situation, he must observe again to see what has changed and what the new circumstances are, and so the loop continues.

This loop shows how decision making, especially in crisis, is a continuous and cyclical process. The leader who can effectively cycle through the loop faster than his opponent is in a tremendously advantageous position. By the time the other side acts, the faster side has already moved on to its next action, so that the slower side's action is no longer appropriate. Over time, the opposing side's actions are wrong by larger margins in a collapsing time frame until they eventually become totally ineffective. The leader who accomplishes this mental agility is said to be inside the other side's OODA Loop, and research suggests that's a winning edge.

Tempo and momentum are vital elements generated by decisiveness. By shortening the time needed to plan, make decisions, communicate, and coordinate those actions, the leader can get inside the OODA Loop of his opponent, improving chances for success. Rather than absolute speed, however, this process should be thought of as relative speed. The goal is to be faster than the opponent, rather than just being quick or decisive as an end game. A small advantage exploited repeatedly is often far more decisive in the end than a much faster cycle of only one or two loops.[232] Ultimately, the key is timely application of meaningful action.

Business organizations and individuals cycle through OODA Loops every day. A product manager discovers that a competitor has created a product in direct competition with the one her company produces. Immediately, she looks at the competitor's production capabilities, observes the consumer environment, and examines the marketing budgets of both products. She thinks about how the new product will effect her own sales. She then decides to implement a new marketing strategy highlighting features the competitor's new product lacks. She watches the market to see if her product can hold its own or if she needs to take additional action. Action has been taken—the OODA Loop completed—and the next cycle begins depending on how the market responds.

The same cycle occurs in more complicated situations, but the same logic applies. Decision makers gather information (observe), form hypotheses about customer activity and the intentions of competitors (orient), pick a course of action (decide), and execute (act). The cycle is repeated continuously. Other decision makers in the same organization also will be moving through their own loops simultaneously, perhaps with a different tempo. The aggressive and conscious application of the process provides an advantage over a competitor who is merely reacting to conditions as they occur or has insufficient situational awareness. Although business loops often are slower than military or combat loops, the leader who has the ability to move through an OODA Loop faster when situations change (like seeing a competitor's move, seeing how international news is going to affect their product, etc.) is going to come out on top.

An additional method of decision making helps to understand how some decisions are made: the identity method. In these circumstances, the decision maker considers how he defines himself or how he would like to define himself. He then thinks, "How would someone like me decide in this kind of situation?" or "How would someone I'd like to be decide?" A person might think of himself as a liberal democrat, for example, and if he were called on to vote on an issue he did not have strong feelings about, he might consider how liberal democrats in general thought about that issue and then vote accordingly.

Translated to the case of a Marine NCO faced with a decision about how to behave or decide in a particular situation, the same guiding principle could be brought to bear. The corporal or sergeant might not know what do, but he or she is familiar with what's generally expected of a Marine NCO and decides to act accordingly.

Combat missions involve rapid and, hopefully, effective decisions. This was true in Sergeant Wren's day, and it remains true today. Unfortunately, piracy remains as well. In September 2010, a group of Marines found themselves facing a situation involving these modern-day brigands.

Take the case of one of the most spectacular and successful operations against Somali pirates, for instance. The USS *Dubuque*, part of the Peleliu Amphibious Ready Group, was supporting maritime security operations with the U.S. 5th Fleet operating in waters of the Persian Gulf,

Red Sea, Arabian Sea, and off the coast off East Africa as far south as Kenya. On the night of September 7, 2010, the Dubuque was attached to Combined Task Force 151—the international counter-piracy task force—when Somali pirates threatened the commercial freighter MV *Magellan Star*.

An outline of the events that followed is best expressed by paraphrasing the narrative from Captain Alexander Martin, commander of the Force Reconnaissance platoon that composed the striking force of the U.S. naval force that responded to distress calls from the Magellan Star.

On the morning of September 8, pirates intercepted the German-owned Magellan Star steaming in the Gulf of Aden and her crew sent out a distress call. A Turkish Navy frigate operating in the area heard the transmission and passed the word to the U.S. flagship that Somali pirates were boarding the merchant vessel and that the crew had locked themselves in the engine room. The closest U.S. Navy vessels, USS *Princeton* and USS *Dubuque*, were ordered to the scene, approximately eighty-five miles southeast of Mukallah, Yemen. It was time for Captain Martin and his Marines to undertake some rapid decision making.

As the *Dubuque*, carrying the reaction force, steamed toward the captive vessel, Martin said his Marines fell into a pre-planned routine that they'd practiced for nearly a year as the 15th Marine Expeditionary Unit's Maritime Raid Force. In just one hour his Marines:

> *"pulled pre-staged shooter's kits, body armor, weapons, ammunition, communication and breaching equipment and moved it to our assembly area. Comm was op-*

checked, weapons were function checked and set in the best
condition possible, shooters performed their pre-assigned
tasks to meet our conditions for a 120-minute alert status
while assistant team leaders conducted simultaneous in-
dividual inspections: flotation devices, chem lights,
breathing devices, roster cards, tourniquets, medical
equipment, lights, night vision, weapons, comm ... all
given one last op-check."[233]

As updated information came in, the alert status was
shortened: sixty minutes, thirty minutes. The Marines were
re-briefed on all pertinent intelligence as they waited for
someone in authority to pull the trigger. Martin and the
Marines waiting aboard the Dubuque were running
through various scenarios that might occur, but there was
another set of decisive planning procedures rolling and
cycling higher in their chain of command. That was based
on a method called the Rapid Response Planning Process
known as R2P2.

R2P2 is the process of organizing and coordinating
short-notice operation plans. When deployed, Marine
Expeditionary Units are designed to be a quick-response
force, so Marine and Navy planners need to create the
plans and be ready to execute them within a six-hour
window. As a part of their pre-deployment cycle, each
MEU receives specific R2P2 training and continues to
practice, with the goal of formulating and executing a plan
in fewer than two hours.[234] This process has been developed
to prompt decisiveness in the planning process on operations where timing is crucial and delays are deadly, as was
the case with the pirate-held Magellan Star in September
2010.

The trigger Martin and his Marines had been watching was tripped shortly after USS Dubuque came within range of the captured vessel. The green light flashed, and the Marine boarding party went into action. During that boarding, Marine Sergeant Richard Weir served as one of the snipers covering the raid force, and he believes it was the training that made the difference in what happened next:

> *"From the arduous and complex planning to the execution, we did it all over and over again. We used different ships and different scenarios and even live fire. But where we really made our money was in our debriefs."* [235]

During those post-game debriefs, the actions and thoughts of everyone involved in an exercise or operation were reviewed, critiqued, and evaluated. The Marines who made the successful boarding of the Magellan Star had a solid understanding of what had worked well, what was not as effective, and what needed improvement. "Everyone came together, enlisted and officer, Navy Captains and Marine Sergeants talking openly about the ways to perfect this dangerous and high-stakes mission," Weir continued.

> *"People often call the Marines old-fashioned, and in a lot of ways we are, but what kept us alive and what keeps us alive is our ability to adapt and progress by way of these open forums."* [236]

In Weir's experienced view, all that training results in what's needed for leaders to be decisive. "Decisiveness comes into play at all levels," he emphasized. "Without decisiveness, nothing is done—or worse yet, the things that you fear most will happen."

U.S. Marine Corps snipers (left to right: Cpl. Y.V. Kulgeyko, Cpl. D.P. Talone, Sgt. M.D. Fechner, Sgt. J.D. Seifert) provide cover and observation during an operation Sept. 9, 2010, to recover the motor vessel Magellan Star from suspected pirates who took control of the ship Sept. 8, 2010. (Department of Defense photo by Cryptological Technician 1st Class Gregory Tate, U.S. Navy/Released.)

R2P2 was in full operation throughout the day, and the plan was briefed and confirmed. But the command to execute did not come that evening. The Marines grabbed what sleep they could before their 0300 reveille. While they slept, the execute order moved up the chain of command, all the way to the President of the United States. The word to go finally came at dawn on September 10. Martin led his Marines into action and credits their success on "the individual actions of twenty-four highly-trained shooters who were put in decision points of the highest moral magnitude: when to shoot, when not to shoot."

Martin is restricted by operational security from describing some details but he is outspoken about the result of their decisive action:

*"... the long and short of it was: some of the enemy threw
their hands up when rifles were put in their face, some
ran and attempted to elude us in the superstructure but
were run down and some hesitated but were taken down
by less than lethal force, as the situation dictated."*[237]

Once Martin's shooters, dubbed Alpha Element, had
subdued the pirates aboard the *Magellan Star*, they communicated with Bravo Element aboard the nearby *Dubuque*
and got some help in communicating the same information
to the ship's crew, which was still locked away in hiding.
Sailors aboard the *Dubuque* used loud-hailers to amplify a
call to the crew asking them to come out of hiding and
letting them know U.S. Marines now had control of their
vessel.

The captain of the *Magellan Star* finally peered cautiously through a small window in a steel bulkhead, but he was
still unconvinced that his ship was back in friendly hands
and that it was safe for his crew to emerge from hiding. At
that point, one of the Marine NCOs in the boarding party
exercised both decisiveness and innovation. Sergeant Max
Chesmore ripped an American flag patch from his shooter's
kit and poked it into the compartment as a bit of reassur-
ance. That did the trick. The captain broke into a huge grin
and the crew emerged behind locked doors.

When the ship was back in proper hands and the count
was tallied, the Marine boarding party had rescued eleven
crew members and captured nine pirates. The confused
Somali hijackers were promptly slapped into custody
aboard the guided-missile cruiser USS *Princeton*.

Lieutenant Colonel Joseph Clearfield, commander of
Battalion Landing Team 1st Battalion, 4th Marines (BLT-

1/4) to which the boarding party was attached, called the mission "a 10 out of 10 on any scale."[238] In the end, the mission's success was all about Marines like Sergeant Matthew Fechner, a scout-sniper with the MEU's ground combat element making good decisions and acting on them without having to delay or dawdle in the process.

Sergeant Fechner spent most of the time behind his sniper rifle that day, scanning the decks of the *Magellan Star* through his rifle scope and contemplating what actions he might have to take if one of the boarding party Marines was threatened by an armed pirate. Precision shooters were positioned in a number of areas to make sure the boarding party's backs were covered. At one point, he observed a pirate walking aboard the ship with what appeared to be a Soviet-made rifle. Fechner had a decision to make.

"Now, the situation can go two ways," he remembered:

"First, I had determined this guy was a possible threat. The Rules of Engagement are telling me that he is not, although if I had a different feeling, I could decide to take the target out. On the other hand, I understood the situation fully and decided that I was going to wait and not take the kill shot. Some may argue that I should have taken that shot, but more would say that I was right in my decision."[239]

The decision in that moment was purely Fechner's, and he made it based on the situation and the potential outcome of pulling the trigger.

Sergeant Fechner was decisive, and he has thought often about that in the time since the encounter with the Somali pirates aboard the *Magellan Star.*

"In those circumstances making a decision, whether good or bad, is better than making no decision at all. There were many decisions like mine that had to be done quickly, yet also very carefully and with the knowledge of what was practiced, planned and briefed to each individual element."[240]

And those decisions, Fechner believes, can only be made properly by the people involved:

"Decisions are usually done best by the guys on the ground. Among the Marine Raid Force, you have your leadership structure, but during critical times, anyone with a good decision based on the developed situation becomes the leader at the time, and takes on the role of being in charge and doing what he sees as the best way to solve and execute the mission at hand."[241]

Sergeant Weir couldn't agree more:

"Decisiveness came into play at all levels here. From my scope to the ensign steering the RHIBs [rigid-hulled inflatable boats] toward the Magellan Star, we all had to evaluate the situation at hand and determine what we were going to do to manipulate the situation in our favor and then execute that plan almost instantly. I will say that prior to the launch of this mission, every key player spent more than a few hours ready to go while we waited for an example of decisiveness (on the part of the U.S. government at the highest levels). Approval for the mission came from very high up the chain of command and it took more time than there was in the first day we arrived on scene. Fortunately this did not affect our ability to exact justice on the pirates."[242]

Decisiveness, coming from a sharp mind that understands how the leadership trait can be applied in real-world situations, is clearly vital for success on the battlefield or in the boardroom. But leaders cannot rely on their intellect alone. Like the body and the mind, a resolute spirit must be engaged for successful leaders to bring their teams to victory.

Part 3

Spirit

Chapter 11

Integrity

"I've made the points that leaders under pressure must keep themselves absolutely clean morally (the relativism of the social sciences will never do). They must lead by example, must be able to implant high-mindedness in their followers, must have competence beyond status, and must have earned their followers' respect by demonstrating integrity."
—Admiral James B. Stockdale (1923–2005)

If there is one constant about the fourteen leadership traits recognized by the Marine Corps, it is that no one of them stands alone or above all the others. They are intertwined and interdependent like the parts of a fine watch or the gears in a complex machine. Absent one trait, all the others are affected: the watch loses time, and the machine malfunctions. Without unselfishness, it may be difficult to be dependable. Without knowledge, it's tough to make solid judgment calls.

Integrity, too, involves elements of the other traits: dependability, judgment, knowledge. It is a quality learned and exercised effectively over time. It is not something applied only to specific situations. It is the result of ongoing experience in dealing with people and making decisions that affect them. It can be tough to define and even harder to practice.

The accepted meaning of the word comes from its Latin root *integer*, meaning something that is whole or intact.

Engineers, for example, define structures with integrity as those that have internal consistency or lack of corruption. While they generally are speaking of steel girders or chunks of computer code, that definition also serves admirably in describing leaders with integrity.

When considered an aspect of leadership, integrity can be said to be a trait that brings a sense of wholeness or completeness to a group. Leaders who demonstrate integrity create an environment of common values that bind people together, and Marine Corps history is shot through with excellent examples.

Carlos N. Hathcock II, one of the Corps' most renowned snipers, had integrity in huge helpings, according to the people who knew him best during his colorful career as a shooter and as an instructor. Born and raised in the backwoods of rural Arkansas, Hathcock enlisted in 1959 when he was seventeen years old. At five feet ten inches in height and a skinny one hundred forty pounds, he was not an imposing recruit. He was slight but strong with a musculature developed when he dropped out of school at fifteen and went to work for a concrete contractor. By the time he enlisted in the Marine Corps after shoveling cement six days a week, he could lift his own weight over his head and run all day long.[243]

Following Marine Corps recruit training at San Diego, California, Hathcock quickly established himself as an expert marksman. He set records on the rifle range and won awards, including the celebrated Wimbledon Cup for marksmanship in 1965.[244] It was a troubled time for America with a growing commitment to combat in Southeast Asia. While the Marine Corps celebrated champion shooters, it couldn't afford to let top shots just continue to complete and win trophies. That kind of marksmanship was better used against targets in Vietnam.

With his acumen in rifles, Hathcock was classified as a sniper and sent to Vietnam in 1966, where he joined a growing team of Marines offered to infantry units as precision shooters. A number of the infantry commanders either didn't know how to use snipers or mistrusted their independent method of operation, but that didn't bother Hathcock, who consistently volunteered for long-range, hazardous missions during the two years he served as a sniper in Vietnam. Within a matter of time, Hatchcock was dubbed "White Feather" by the Viet Cong and North Vietnamese enemies who had come to know and fear him. An article about him appeared in the Los Angeles Times, quoting a commanding officer as having to restrict Hathcock to quarters to make him rest. Hathcock didn't have much patience with resting, and it was a sentiment he often repeated: "It was the stalk that I enjoyed," he said in an interview with the *Washington Post* in which he described the art of the sniper in combat:

> *"Pitting yourself against another human being. There was no second place in Vietnam—second place was a body bag. Everybody was scared and those that weren't are liars. But you can let that work for you. It makes you more alert, keener, and that's how it got for me. It made me be the best."*[245]

By the time he left Vietnam in 1969, Hathcock had proved the value of snipers to the combat effort, and the Corps was anxious to use his expertise to establish a formal sniper training program. It was time to formalize the business of sniping, which had mostly been an ignored or dormant skill between wars when excellent marksmen just had to wing it. Hathcock helped establish what has become the U.S. Marine Corps sniper training program at Quantico, Virginia, a course that has become the most prestigious school for military shooters throughout the world.

Hathcock wanted to bring integrity to snipers, to bring groups of outstanding marksmen and reliable stalkers together with simple, singular values and training. He did it in the early days of the sniper program by sharing his views about the value of continual training and personal skill development, straightforward and honest speech, and a pride in their mission which had come to be defined for the first time in official manuals: "To support combat operations by delivering precise fire on selected targets from concealed positions."[246] That's simple and straight forward enough, but Hathcock also assured his students that there was more to it than that and that much more value was added by effective shooting and field-craft. Snipers have a profound impact on an enemy's psychology, he told his students. They create enemy casualties as a matter of course, but they also sow fear, slow down enemies, frighten and demoralize them, and foment confusion, which adds an extra advantage to Marines on the more conventional firing lines.

Charles "Bill" Henderson, a Marine Chief Warrant Officer, was a long-time friend who eventually became Hathcock's biographer. He watched the effect Hathcock's legacy had on Marine snipers who followed in the footsteps of White Feather:

"When you're operating, typically a sniper team is two men. When I'm talking about anything to do with snipers, mind you, it means it came from Carlos. Carlos Hathcock established sniper doctrine as it is today. The Marine Corps, they saw Carlos, they examined what he did, and why he was successful. They looked at those attributes and that's what they sought to repeat when making other snipers. He was kind of the pattern."[247]

Even today, Hathcock's name is well known wherever snipers are training. After retiring from the Marine Corps in 1975, he worked diligently to develop snipers within the community of law enforcement officers.[248] Though he was known for many years for his impressive number of ninety-three confirmed kills in Vietnam, including one Viet Cong shot dead at the unbelievable range of 2,500 yards, he took no pleasure in the numbers. "I'll never look at it like this was some sort of shooting match, where the man with the most kills wins the gold medal," he once said.[249]

Incredibly, Hathcock won only one decoration for valor and it had nothing to do with sniping. In fact, the action ended his career as an active shooter. On September 16, 1969, Hathcock was riding in an amtrac (amphibian tractor) on Vietnam's infamous Route 1 when the vehicle struck an anti-tank mine. He was briefly knocked unconscious, sprayed with flaming gasoline, and thrown clear. When he awoke, he pulled seven Marines off the burning vehicle before jumping to safety. It took a while for Hathcock to be recognized for his actions, but he finally was awarded the Silver Star in 1996. According to his citation, he acted "with complete disregard for his own safety and while suffering an excruciating pain from his

burns, he bravely ran back through the flames and exploding ammunition to ensure that no Marines had been left
behind." His days as a sniper and as a competitive shooter
were over due to the burns he suffered in that incident.
Before he was finally discharged from a series of hospitals,
Hathcock had endured thirteen skin graft operations.[250]

In Henderson's book *Marine Sniper: 93 Confirmed Kills*
about Hathcock's career, there's a quote that goes beyond
the service records: "I like shooting, and I love hunting. But
I never did enjoy killing anybody. It's my job. If I don't get
those bastards, then they're gonna kill a lot of these kids
dressed up like Marines. That's the way I look at it."[251]

"Carlos was all about supporting the infantry rifleman,"
Henderson wrote. "And his integrity was such that his
actions supported what he said. Everything he did was with
mission first, then support your Marines."

Washington Post reporter Stephen Hunter, the author of
several novels featuring snipers, said a sniper is not only at
"the point of the spear, he *is* the point of the spear."
Hathcock would have hated that. To him, Marine riflemen
were the point of the spear and the purpose of having one
in the first place. Snipers were on the battlefield to support
infantrymen and for no other reason in Hathcock's mind. It
was a matter of integrity for the master sniper, and
Hathcock felt a need to teach others, both superiors and
subordinates, to view snipers in that same support role. It
was a hard sell for a man who had become as famous in
military circles as he had. Henderson knew that top shooters like Hathcock got a heady bunch of notoriety, but that
they didn't really seek that type of renown:

"The sniper doesn't seek it, but it's one of those gee-wiz jobs. Kinda like recon, you know, a guy who jumps out of an aircraft with a knife in his teeth. Carlos honestly believed that every Marine's job was equally important, and he said, 'Look at that admin guy. Look at the guy working in disbursing. Their jobs are just as important as mine. They're different, but they're just as important. If they didn't do their jobs, I couldn't do my job. Everything is in support of the infantry. Why do we have the air wing? To support the guy on the ground.'"

In a practical fashion, Hathcock put his convictions on display. He went out of his way to visit the cooks and bakers, Marines working in the supply warehouses, and those toiling in administration offices during his periodic visits to bases following discharge. "He would go out of his way to go visit with those guys," Henderson recalled, "and tell them how important they were."

Hathcock was an honest man and an example of how that admirable quality is an integral part of integrity. He knew what he was and what he'd done, and no amount of publicity or notoriety could compel him to embellish that. He didn't try to reinvent himself or be someone he was not. He was a plain speaker of the truth as he saw it, as Hunter pointed out in his article comparing the man with the legend:

"In all the endless revising done in the wake of our second-place finish in the Southeast Asia war games, he never euphemized, didn't call himself an 'enemy troop-strength reduction technician' or 'counter-morale special-

*ist.' He never walked away from who he'd been and what
he'd done."*[252]

Hathcock remained a salty but self-effacing Marine
NCO, even from a wheelchair after a combination of
multiple sclerosis and the burns he had received over 43
percent of his body during the amtrac incident in Vietnam
made it difficult for him to travel or teach.

"In my mind," Henderson said, "Carlos was the greatest
sniper of the Vietnam era, not because of the number of
kills, but because of the quality of the missions. The
missions he went on were daring." One of the most famous
was his hit on a Vietnamese general. With trademark
patience, Hathcock crawled about a thousand yards across
an open field. It took him four days to get into a position
where he could get a shot and remain under some semblance of cover, and there he waited, motionless. Enemy
patrols passed by, walking with dogs, which increased the
likelihood of Hathcock being detected. He went undiscov-
ered, surviving on one canteen of water, when he dared to
drink.[253] He successfully completed that mission, which
culminated with an incredible 700-yard shot that dropped a
very high-value enemy target with one round. Henderson
noted that dropping the mission wasn't an option for
Hathcock:

> *"Crawling days and nights over open fields to get that
> General..., and he knew that he probably wouldn't come
> out alive. But he made his shot and he did get out alive.
> He told me, 'I could have gone home. But then they would
> have sent someone else, they would have had less of a
> chance than I did. I had to take that mission.'"*[254]

It was a deep and abiding sense of integrity that led Hathcock to continue under such circumstances, even when he had an easy chance to go home and rest on hard-won laurels.

Lt. Gen. P. K. Van Riper, Commanding General Marine Corps Combat Development Command, congratulates Gunnery Sgt. Carlos Hathcock (Ret.) after presenting him the Silver Star. Standing next to Gunnery Sgt. Hathcock is his son, Staff Sgt. Carlos Hathcock, Jr. (Department of Defense photo by Sgt. James Harbour U.S. Marine Corps/Released.)

Integrity in a leader is reflected by honesty as well as by a desire to inspire and a devotion of values that the leader constantly tries to communicate to those he or she leads. The leader with integrity can rarely if ever relax a commitment to what he or she believes is the behavior that best reflects those close-held values. When followers see leaders acting with integrity, they are more likely to want to emulate that quality. Conversely when subordinates see a lack of integrity in a leader, they feel far less obligation to follow appropriate behavior patterns themselves. Hathcock learned this first-hand when he returned to Vietnam on a second tour.

Upon arrival in Vietnam, Hathcock was immediately assigned as the new leader of the 7th Marines scout-sniper platoon. He did not like what he saw when he headed down the hill from the 7th Marines command post to the sniper platoon's hooch to assume his new responsibilities. The place was familiar to him; he had helped build the sniper's enclave two and a half years earlier on his first tour in Vietnam, but he quickly discovered that much had changed.

The whole area seemed deserted until he found the platoon sergeant drinking beer in a filthy, trash-filled room. Hathcock introduced himself as the man's replacement, took note of the filthy surroundings, and asked if the sniper NCO had just returned from a mission in the bush. It was the only reason he could think of that would explain why things were so disorderly. The sniper sergeant told Hathcock that he was not only not just out of the bush but that he rarely ever went there, as the command had no idea what to do with snipers. For the most part, the NCO told his replacement, snipers were used as handymen around the command post doing odd jobs like cleaning latrines and standing perimeter security. He had no idea how many were in his platoon or where they might be at the moment.

Hathcock was more than a little insulted by the lack of unit integrity and a similar lack of that quality on the part of the man he was replacing. These misused and badly led Marines were now his snipers, and it was clear Hathcock had a serious leadership challenge on his hands. He booted that sergeant out of the hooch and told him to report to the sergeant major for reassignment. He then spotted one of the members of his platoon and told him to gather the others with their sniper gear and meet in an hour. The messenger passed the word in a hurry. The 7th Marines snipers had a new leader, and things were bound to change for the better. They were about to get a healthy dose of unit integrity from a leader who understood the concept.

In short order, all the Marines assembled, albeit in raggedy clothing. Hathcock told them that from then on, they would dress in uniform and look like Marines again. The next day, they would begin training, to zero in all the rifles, and be certain that they were ready to go out on missions. For the rest of that day, they were to clean up the trash and filth. The sniper hooch was to be a real headquarters again, with Marines standing watch. No longer would it be a lounge or playpen.[255]

The situation Hathcock encountered in the sniper platoon was a result of a circular problem not uncommon in military circles. Because the command element did not know how to effectively deploy snipers, they were assigned routine tasks that did not make use of their specialized skills. And because the snipers were not respected or properly used, they lost motivation, which made them lazy and counter-productive when they should have been vital combat assets. That merely reinforced opinions among senior leaders that the snipers were not worth much as Marines and therefore could not effectively support the overall missions. Snipers were rarely called on to do what they were trained to do, so they started to believe that what they did wasn't worth much—and neither were they. It was a deadly and devastating loop that Hathcock was determined to break.

He did that very quickly by demanding that the snipers behave as professionals no matter what they were assigned to do. The snipers in his platoon reacted favorably because they recognized that a leader with integrity was now in command, and they could almost feel the pride—the unit integrity—begin to improve. Soon enough, they looked and felt like competent Marines, and the leadership above Hathcock started looking at the snipers from a new perspective. Hathcock made certain that his snipers were ready to go out on missions, and trained those above him by informing his superiors as to the best way to deploy the men, who turned around completely when they finally were entrusted to do their jobs.

That didn't please everyone in the command. Other NCOs warned Hathcock that there would be resentment now that the snipers were no longer available to do routine dirty work that would have to be shared among other Marines. Hathcock had little time or empathy for such complaints, and he had no fear of becoming unpopular among his fellow NCOs:

> *"My snipers come ahead of my own pleasure. I've got a hunch that fella I just relieved might have tried to get along and keep everybody happy. He was a good Marine when I first knew him. But you can only compromise so far. I think he chose to keep folks on the hill (senior commander) happy, and his Marines went to hell in a hand basket."*[256]

Hathcock was not willing to compromise his integrity for anyone, and he certainly wasn't willing to do so at the expense of his Marines. He found a solution that allowed him to keep his integrity by exercising his own judgment based on the knowledge and training he had accumulated as a Marine.

Problems similar to what Hathcock encountered in Vietnam are not unique to the Marines. In fact, they are just as common in other pursuits, and civilian industries have come to note that Marines have a particular talent for solving them.

During the economic boom of the 1980s, when the American economy had the longest period of sustained growth during peacetime in U.S. history,[257] there was a period when American industry looked for leadership outcomes similar to those of the Marine Corps. Companies claimed to want people to be motivated, to be happy to come to work, to have a family-type bond among each other. However, when the economy began to sour, more and more industries became more interested in bolstering the short-term bottom line. Looking for ways to cut corners on cost, product quality became less important than product profit. Downsizing became a common means of lowering costs.

That tactic became commonplace, effecting scores of industries. For example, after the attacks on September 11, 2001, the airline industry was filled with uncertainty. The industry did not know how those attacks would affect regulations or the number of customers willing to fly, and the country feeling the first pangs of a recession. Every airline company in the United States responded by announcing tens of thousands of layoffs—except one: Southwest Airlines.

"We've never had layoffs," Southwest founder and former CEO Herb Kelleher declared. "Our people know that if they are sick, we will take care of them. If there are occasions or grief or joy, we will be there with them. We value them as people, not just cogs in a machine."[258] Kelleher was displaying both individual integrity and an appreciation of unit integrity in his response to the crisis.

As of 2010, Southwest could claim to be America's largest domestic airline (based on domestic passengers carried), and it has maintained a market capitalization larger than all its domestic competitors combined. Downsizing is not part of Southwest's culture. Its former head of human resources said, "If people are your most important assets, why would you get rid of them?" Why indeed if the company's integrity is measured at least in part by the strength, loyalty, and motivation of its employees?

Although Southwest avoided downsizing, other airlines—as well as a number of other industries—have instead chosen to layoff or reassign workers, typically in an attempt to beef up the bottom line for the short term. This often results in overworked, unhappy workers who may feel that their skills are not being used to the best advantage—not unlike what Sergeant Carlos Hathcock faced with the snipers of the 7th Marines in Vietnam. It is a pattern often repeated.

As a short-term response to economic uncertainty, companies begin to lay off employees. That translates to trained workers—whether Hathcock's snipers or airline employees—being unemployed or forced to work outside their area of expertise. With a smaller and often demoralized staff, productivity and customer service among airlines drops, just as sniper performance had decayed before Hathcock returned to Vietnam and the 7th Marines. That same deadly, self-defeating loop begins to close.

Of course, airlines and sniper platoons aren't the only examples of what can happen when integrity is not an integral part of leadership. The cycle of doom also struck popular electronics retailer Circuit City. To cut costs and bolster profit margins, Circuit City eliminated 3,400 people from its sales staff in 2007, leaving fewer people to make sales and care for customers who were increasingly unhappy with their treatment. Competitors saw an opening and moved to take advantage, causing Circuit City to respond with another round of service-debilitating layoffs. All integrity was lost, the loop closed, and Circuit City filed for bankruptcy in 2008.[259] A little integrity in leadership and corporate response might have made for a different ending.

Good leaders like Carlos Hathcock and Herb Kelleher understand that. They also understand that people can't be effectively coerced into supportive, successful performance by threats of punishment or loss of employment. It doesn't work that way in units led by people with real integrity.

Having identified integrity as a current or potential problem in an organization, whether it's a multi-million dollar company or a Marine Corps rifle squad, how can a leader establish or increase integrity? Rewards, free communication of values and ideals, clear expectations and demands, and proper examples of leadership are all dependable tools worth considering, but each has to be dealt with carefully and with integrity. For example, if rewards go only to those with high performance, such as top sales numbers, that creates an environment where numbers are everything and the message is to get them at all costs. The pressure to make the numbers can create an environment that suffers from a lack of integrity.

Building a culture of integrity is crucial to success. Shared principles and shared practices lead to a culture of integrity, but they must be created and supported by effective leaders. In getting there, the leader has to ponder a number of questions: What are the values that the organization ... well, values? What practices support those shared values? What is the benefit for an individual who follows the practices?

At Southwest Airlines, Kelleher has made a practice of leading with integrity. In 2001, when facing a financial crisis in an organization where the corporate culture demanded avoiding involuntary layoffs, he opened the communication channels and asked his employees for ideas about how to shave costs. Kelleher was searching for effective practices that supported the values of his company while improving their practices and bolstering the bottom-line. He found them. Within six weeks, he had received feedback from his employees, whose recommendations totaled billions of dollars in savings.[260]

Of course, it's not always that simple. It is not possible to completely eradicate fraud, corruption, or human error that can lead to a loss of unit integrity. Organizations and institutions are made of human beings, with all of their attendant flaws and differing viewpoints. Limiting damage to personal or unit integrity caused by such failings is generally a matter of damage control. When bad things happen, the reaction to them often demonstrates integrity or a lack of it. Both civilian and military organizations are shot-through with good and bad examples of integrity gone wrong.

For example, on December 8, 2008, Marine aviator First Lieutenant Dan Neubauer, at the controls of an F/A-18 Hornet, took off from the aircraft carrier USS *Abraham Lincoln* off the coast of southern California on a routine training flight. After experiencing problems, he shut down one engine and turned toward Marine Corps Air Station Miramar, hoping to get his aircraft on the ground before anything more threatening happened. He was just three miles from the Miramar runway when his second engine flamed out and his electrical system failed. Neubauer ejected from the doomed airplane. He held on until the last possible moment, but after he punched out, his crippled jet crashed in a populated area, killing four civilians from the same family.

The Marine Corps launched an immediate investigation and discovered that some critical decisions made during the incident didn't reflect very well on procedures and practices. The Marine Corps could have tried to protect its reputation and kept its finding in-house, but there was integrity at stake, so they opened the books and told the public what they found. Major General Randolph Alles, assistant commander of the 3rd Marine Aircraft Wing, said the crash was:

> *"'clearly avoidable,' the result of 'a chain of wrong decisions.' Mechanics had known since July of a glitch in the*

jet's fuel-transfer system; the Hornet should have been removed from service and fixed, and was not. The young pilot failed to read the safety checklist. He relied on guidance from Marines at Miramar who did not have complete knowledge or understanding of his situation. He should have been ordered to land at North Island."[261]

The fall-out was swift, public, and comprehensive—and a classic demonstration of organizational integrity. As a result of the investigation, twelve Marines were disciplined; four senior officers, including the squadron commander, were relieved from duty. The organizational openness had a supportive effect on public opinion of the Marine Corps.

Bob Johnson, who lived behind the Korean family and barely escaped himself, said, "The Marines aren't trying to hide from it or duck it. They took it on the chin."[262]

The Marine Corps strives to be an organization that holds itself accountable for its mistakes and shortcomings, and that's integrity at the highest levels. In this case and others like it, their transparency is also an example of the difference between public relations damage control and an honest approach to responsibility and integrity. It did not go unnoticed. In a *Harvard Business Review* article on the topic of organizational integrity, Maura Sullivan, a former Marine Captain and MBA/MPA candidate at Harvard's Business and Kennedy Schools, wrote:

"I was not surprised by the actions taken by the Marine Corps. Integrity is the most valued Marine Corps leadership trait; It is absolutely central to the Corps' identity and is paramount to any other attribute. Consequently,

the Marine Corps training and culture is explicitly designed to foster an adherence to such."[263]

Integrity at the very top often engenders similar displays at much lower levels. Sergeant Sean Bunch, a Marine from 1994 through 2006, has a vivid memory of an integrity violation that helped him evolve as a leader. In 1995, he was working with Joint Task Force 6, a Department of Defense multi-service operation for counterdrug and anti-terrorist operations, on the border of Mexico. The Marines were carrying live rounds when they ran their missions so safety and judgment were critical issues for everyone involved. Bunch was approached by a Marine who had noticed that a crucial retaining pin in his rifle had been replaced with a toothpick. It was a dangerous situation that could have caused a weapon malfunction. Task Force NCOs were determined to find out what happened. "Some people thought he might have done it to his own weapon to cover the fact that he'd lost a piece from his weapon," Bunch recalled during an interview. "Our sergeants were determined to get to the bottom of it."

Everyone who was questioned denied knowing how a toothpick had come to be inserted in the weapon, and for two days operations ceased while the investigation continued. After denying any involvement several times, a corporal finally admitted his guilt. "This guy lost so much respect on so many levels he was never looked at or taken seriously again," Bunch continued:

"The problem was that he lost his integrity. Putting a toothpick in a weapon as a retaining pin was extremely horrifying, especially since we were running real world

missions. But even worse was his straight-up denial during the interrogations over the previous two days. He let all the rest of us get tore to pieces when he could have stopped it at any moment."

Bunch had witnessed a practical example of what integrity—or the lack of it on the part of an individual—can do to an outfit:

> *"What I learned from this is if he just would have had some integrity and admitted his mistake right away, he might have had a chance of actually keeping some respect from all of us. I learned a lot from seeing how stupid this corporal looked after it was all said and done. During my career, if I made a mistake, I learned it was better to roger up to it and carry on instead of trying to hide it or deny it."*[264]

Bunch came away so impressed by the lesson that when he became an NCO himself. he made extra efforts to teach the value of integrity to his Marines.

Integrity and responsibility are taught from the first day of recruit or officer training, and that education continues throughout a Marine's career. Integrity violations are taken seriously, and response to these violations is swift, as it was in the case of junior officers who were booted out of training—and out of the Marine Corps—in 2010.

In May of that year, thirteen Marines were discovered cheating on a land navigation exam while in training at The Basic School at Quantico, the basic-level course of instruction taken by all Marine officers. The Marine Corps command was swift, honest and open in its response. "The

Commandant has made it clear that we can tolerate many things, but not integrity violations," said Lieutenant Colonel Matthew McLaughlin, a Marine spokesman. "Personal integrity is the heart of Marine Corps leadership."[265]

It's not only officers who are held to such a high standard. Today's drill instructors know all about those standards. One of these top instructors, working out of Marine Corps Recruit Depot Parris Island in South Carolina, recalled a case in which he had to make a judgment about a young man's suitability to become a Marine that hinged on more than simple performance:

> *"We had a boot that failed to qualify on the range. Now, the Drill Instructors are watching closely all throughout training, and they know both the recruits and the equipment. This guy bent his rear rifle sight, and then said it was that way when he drew it, and he needed to requalify with a serviceable rifle. He was booted very quickly."*

That's an example of a lack of integrity that leaders like Carlos Hathcock would not have tolerated in his sniper platoon. According to biographer Henderson, there's no question that the master Marine sniper would have made a similar decision concerning the fate of that young man attempting to qualify as a Marine. Henderson says he learned to value integrity from his friend and from a long, varied career as a Marine:

> *"Integrity is the cornerstone of leadership. In my opinion, that's the only thing I use to evaluate someone's value. Don't care about what car they drive, how much money*

they have in the bank. There's only one thing that no one can take away from you. They can take your life, they can take your savings, they can take your property, they can take everything you've got ... but the one thing no one can take from you is your integrity, your honor. You have to voluntarily give that up. You're the owner of your integrity. And some people sell it awfully cheap.[266]

Chapter 12

Justice

"There is no calamity which a great nation can invite which equals that which follows a supine submission to wrong and injustice and the consequent loss of national self-respect and honor, beneath which are shielded and defended a people's safety and greatness."
—President Grover Cleveland (1837-1908)

Justice is most commonly understood as fairness or impartiality. It's an ancient and cross-cultural concept, so revered and desirable that it forms the basis for most secular court systems around the world. People inherently understand and expect justice in practically all their dealings, from international relations to the price of groceries to the negotiation between buyer and seller at a swap meet. When any of those same people don a military uniform and join an outfit where orders and compliance are rarely open to negotiation, they don't simply leave the concept of justice behind. They may be committed to service in an outfit that is by nature autocratic, but they still expect a square deal. And making sure their teams get a square deal is up to leaders at virtually every level of command.

The tenets of justice seem relatively straight forward. Don't play favorites. Spread the work and the rewards equally. Avoid prejudices or pre-conceptions. Keep performance standards high and insist that everyone meets them across the board. Do all that and you've got an effective unit with high morale. Or is it really that simple?

In practice, justice can be difficult and complicated. For military leaders, justice applied to subordinates may depend on a number of variables over which the leader has no effective control. Considerations such as the length of service, standard duty performance, and the authority a given leader has to enact and enforce justice may come into play. Given what often becomes a slippery slope in considering a very important leadership trait, it's no surprise that practitioners and scholars alike give it some serious attention.

Sergeant Alfred Saucedo, a Marine Reservist assigned to Company G, 2nd Battalion, 24th Marines, is a doctoral candidate in political theory at the University of Chicago researching leadership. To him, "...justice is an indispensable element of small unit combat effectiveness."[267] When Marines believe that:

> *"promotions result from favoritism .... military decorations are awarded on the basis of unit politics rather than personal excellence, and [those] who constantly see infractions of regulations go unpunished are likely to view the Marine Corps and their leaders with cynicism."*[268]

On the other hand, if Marines perceive that they are being treated justly, if they believe their unit is run in a manner that complies with their concept of justice, and if they see others being rewarded and punished appropriately, they have the tendency to follow suit and do what's right.

Staff Sergeant Aubrey L. McDade, Jr., is a veteran Marine NCO who wears a Navy Cross awarded for heroism in Iraq where he was a machinegun squad leader with 1st Battalion, 8th Marines. He has seen it applied—and misapplied—over a varied career that includes service as a Drill Instructor:

> *"When you are dealing with Marines, justice is more a two-sided coin. Let's say you got a Marine that you know that Marine shows up to work every day, he's squared away from head to toe, you can obviously see he put effort into his uniform, he has a high level of fitness, he's very knowledgeable about his job, he's motivated and has initiative, and he's always volunteering, it would be just to recognize this Marine in front of a platoon, and/or company for a number of reasons. To let him know that his efforts have been recognized, he's not just a number, and his work has not been in vain, it would motivate his peer group to be like 'OK, hey, Marine James, this guy's doing this, he's doing that, I can do the same thing, I can get the same recognition for that. I can do it better.'"[269]*

Like many of his fellow Marines, McDade had to travel a long, hard road in and out of uniform to reach a position of leadership, and his journey helped develop his concept of justice. He grew up in tough neighborhoods in Dallas and Fort Worth, Texas, and was the first among his family and friends to graduate from high school. His father was killed by the Fort Worth Police Department when he was eleven years old. It was tough for a young man in those circumstances to see any justice. Raised by his mom, McDade, his brother, and his sister made their way through tough times.

Life often just didn't seem fair, and it was hard for a young man like McDade to catch a break.

Although he tried to apply to college and look for scholarships, none of the colleges seemed interested in him. In fact, few people around him went to college. McDade soon gave up on the idea of higher education. He ran into a Marine recruiter and immediately saw the potential for getting out of rut and finding a way to garner reward for determined effort. That was what the Marine Corps promised, and Aubrey McDade believed he might be able to make his own justice in a life that otherwise was going nowhere.

McDade underwent entry-level training at the Marine Corps Recruit Depot San Diego, and got his first taste of justice as practiced and applied by Marines. It was an experience he carried with him after he reenlisted and became a Drill Instructor himself at MCRD Parris Island in 2007:

"As a drill instructor, justice is one of the leadership traits, and we teach the recruits that. But in our dealings with the recruits, justice does not always seem just. It's more of a tool that we use to get them to bond together as a unit to make them cooperate to make them more elite as a platoon. Justice in a squad bay is different. Justice when dealing with a recruit is not the same as justice when you're dealing with a Marine. Let's say, in a squad bay you have rack mates. They have a certain time to be on line they have a certain time to follow the instructor. Let's say the rack mate of Recruit James is moving too slowly, and he's not responding to the instructions that have been given, but everyone else in the squad bay has. At certain

*times in the training cycle, recruits become rebellious and
they start to pull back because the remedial training sections or the limited disciplinary tools that you have are
not really functioning properly, not really working.*"[270]

Drill Instructors like McDade are taught in their training how to deal with such situations, and he believes that
what may seem like injustice during this initial training
pays dividends for recruits in the long run:

*"So, there are times when, OK, since you did this, and
your rack mate was over there with you, and he didn't
help you, we're going to punish the rack mate also. And
that's not fair or just for that rack mate, but it does build
a unit cohesion between those two, to where that rack
mate is now looking out for his buddy and making sure he
doesn't get into any more trouble. And, his buddy is mad
because he got his rack mate into trouble, so it's a two-
way street with the assistance of each other. But it is, in
fact, justice in the long-term, because these recruits need to
learn to rely on each other to survive in combat, and it
would certainly be unjust for them to leave recruit training without the skills they will need to survive."*

That's pretty much the school solution for Drill Instructors, and McDade used it to great success during his tour
training and molding young men into Marines. It paid
significant dividends, according to Crystal Moore, whose
son graduated as one of Sergeant McDade's recruits:

*"When the newly minted Marines were given free time
after graduation, I had the extreme pleasure of meeting
Sergeant McDade, shaking his hand and thanking him
for his service and for training my son to be the best he*

can be as a person and as a United States Marine. My son told multiple stories about Sergeant McDade and his ability to lead and inspire the recruits who were fortunate enough to have him as a drill instructor. I only wish his story ... could be shouted far and wide to let the American people know what great Marines are serving our country."[271]

Accolades like that serve to bolster McDade's belief that leadership can be taught best by example, but that conviction wasn't simply a result of his service as a Drill Instructor. He came by it the hard way on the bloody streets of Fallujah, Iraq, where he earned a Navy Cross, the nation's second highest decoration for heroism in combat. Fate and an act of international terrorism played a role in getting him to that point.

U.S. Marines from 1st Battalion, 8th Marines, 1st Marine Division prepare to step off on a patrol through the city of Fallujah, Iraq, to clear the city of insurgent activity and weapons caches on Nov. 26, 2004. The Marines are (from left to right) Staff Sgt. Eric Brown, Sgt. Aubrey McDade, Cpl. Steven Archibald, and Lance Cpl. Robert Coburn. (Department of Defense photo by Staff Sgt. Jonathan C. Knauth, U.S. Marine Corps/Released.)

When McDade enlisted in November 1999, he intended to just do his four years. The Marine Corps had proven to be a good fit for him, but his family needed him at home back in Ft. Worth. And then came the world-shaking attacks on the United States on September 11, 2001, followed by American involvement in Iraq. McDade was torn between the needs of his family at home and what had become his extended family in the Marine Corps. As a young NCO at that point, he felt responsible for his Marines and couldn't bear the thought of watching them deploy to combat without him:

> *"I had some new Marines, and I had like twelve Marines that had little training, and I am my brother's keeper. That's in my veins from civilian to Marine, so I volunteered to go to Iraq with them."*[272]

McDade's first combat tour to was March through October 2003, and he spent most of that time operating with his unit near Mosul, about 250 miles northwest of Baghdad. He and his Marines were on deck for twenty-two days. It was a tough time for all the Marines in the combat zone, and there were those in McDade's outfit and in others who thought they were getting a raw deal. McDade saw it differently. It was combat, and that required a different interpretation of justice. Things may seem unjust at first, he said, but justice involves longer thought as well as consideration of issues that go beyond the immediate moment.

When a unit is in training, justice may look quite different from justice in combat. McDade carried that belief home after his first combat tour and applied it in training recruits but he was not through with combat. After completing his tour as a Drill Instructor, McDade returned to Iraq as a machinegun squad leader and found himself involved in the brutal fighting involved in Operation *Phantom Fury*, the Marine Corps' bloody battle to wrest control of Fallujah from heavily armed and firmly entrenched insurgents in November, 2004. It was a pivotal experience for McDade and an inspiration to the other Marines involved in that battle. McDade took the lessons he learned and applied them throughout his career and into future conflicts.

On the night of January 19, 2005, McDade and his unit assaulting Fallujah came under heavy enemy fire, wounding three Marines. McDade sprang into action and brought his machinegun to bear against the enemy. That helped ease the pressure on his pinned down unit, but three critically wounded men were still exposed in an open area that was being pounded by incoming fire. A rescue effort looked bloody if not suicidal for anyone who tried it. McDade decided it didn't matter. Marines were bleeding out in front of his machinegun position and they'd surely die if he didn't do something.

McDade charged into the enemy kill-zone to retrieve the wounded men and pull them to safety. In fact, he did it three times, despite accurate and deadly fire insurgents who targeted him each time he moved. When he miraculously rescued all the wounded, McDade directed efforts to treat the wounded and get them evacuated from the area. His effort on that bloody day in Iraq earned him the Navy Cross but more importantly in his view it saved the lives of two of the three wounded Marines.

It was typical of what was becoming a pattern of inspirational leadership, but McDade says he wasn't looking for recognition. "I take a lot of pride and joy in what I do," he said. "I would give 110 percent with or without this award."[273]

While he's justifiably proud of the prestigious decoration he eventually received during a parade involving graduating recruits while he was serving as a Drill Instructor, McDade believes the decoration had something to say to the newly minted Marines about justice. The recognition:

> *"teaches the recruits how grateful the Marine Corps is [for what Marines do], and hopefully I will be an inspiration to other Marines. I got [the Navy Cross] for the Marines who have fallen and for all the Marines who have done great things and never been recognized."*

In McDade's mind a certain kind of justice was served by people who saw a deserving Marine recognized, and that kind of thing proves that heroic efforts and outstanding performance will be recognized and not ignored.

McDade is enough of a realist and a veteran leader to understand that not all Marines who do great things are recognized, and he knows that doesn't seem fair. It's an understandable reaction among human beings who have a tendency to equate justice with fairness. They seem the same but there's often a big difference. If it is time for chow, everyone in the unit should get the same food in the same amounts, for instance. That's fair, and so it's just. Or is it?

In the case of chow time within a military unit, it may not be just to give everyone equal amounts, even if it's fair.

There may be variables that influence a decision to give some Marines things that others are denied. If certain people in the unit are struggling with weight, it may be fair to give them the same portions, but it's not just considering their efforts to lose extra pounds and comply with standards. That may not seem fair to hungry Marines, but to a leader it's just; it's the right thing to do on their behalf.

Humans have a keen sense of fairness, and it helps good leaders to get a handle on the concept when dealing with subordinates. There's a whole science called "queue theory" that's used in planning for human reaction to the inevitable long lines that usually form in places like banks and theme parks, or where Marines stand in line to receive issued equipment, for weapons inspection, or to board aircraft. The theory is that people don't mind standing in line and waiting as long as they believe the wait is fair to everyone involved. It's a common and easily observable phenomenon.

If there's an adjacent line that's moving more quickly and people in that line are getting served in a shorter span, there's either a resentment or a rush to get in the faster moving line. That's usually counter-productive as the fast line now crowded with new people rapidly slows to a crawl, but its human nature to seek a fair shake. It's also the cause of rumblings, resentment or violent reaction when someone tries to cut in line or slip forward. The purpose of such misbehavior may seem justified in the mind of the line-cutter, but it's viewed as obscenely unfair by everyone else in the line.

Justice is a leadership trait. Fairness isn't, though it can be a useful yardstick in helping to judge what justice needs to be done in many situations. A good leader uses the common concept of fairness to make decisions that need to be made in a just manner. A key to ensuring justice is an understanding on the part of the leader about the perspective of the people he or she leads. It's easier to make a just decision about allotting food, for example, if you know the health of your team, their weight and metabolism, and any particularly difficult duty they may have performed that day.

Of course, leaders can't know every intimate detail about those they lead, and there are times when nothing much more than gut-level instinct and the organization's regulations must serve as a guide to ensuring justice. Unlike their civilian counterparts in industry or other aspects of the business world, Marine leaders have a responsibility for their people on—and off—the job. That creates a leadership burden that challenges the application of justice. Is what a Marine does off duty as important or reflective of service values as what she does on duty? And if off-duty behavior reflects badly on the unit, is it fair that a Marine is reprimanded, prosecuted or punished for it? Staff Sergeant McDade believes punishment may well be just:

> *"Every time we get off for a weekend, or for a 96 [four day liberty] or anything like that, we tell them, 'Don't drink and drive.' We have classes, we have PowerPoint presentations, sometimes we even have spokesmen come through and tell them about drinking and driving, about drugs, and how all these things can ruin your life and your career. Sometimes you have the Marine that goes out, and takes what we like to call a calculated risk. He goes out, he knows it's wrong, but he takes his chances. What are the chances of me getting caught? He rules that*

out and does it anyway. Then, unluckily for him, he gets
caught. As his leader, we have to hold him accountable.
We have to show that if you do good, you get good things.
If you do bad, you will be held accountable."[274]

That might not seem fair, especially to the culprit who thought his personal time away from his leaders was his own affair, but it's a form of justice. It's a difficult concept for civilians to grasp, but Marine leaders understand that what subordinates do during off-duty hours can affect both morale and mission. A Marine who consistently gets away with bad off-duty behavior, such as taking drugs, drinking to excess, or otherwise violating the standard for members of the unit, demonstrates that the situation in that unit is neither fair nor just if everyone else is held to account. Good leaders can't abide that variety of justice.

McDade recognizes that responsibility for administering justice in such situations rests with a unit's commanding officer, but NCOs—those leaders most likely to observe and document misbehavior—play a big role in the military system of justice. In fact, their opinion often is sought when determining both the severity of the violation and the nature of the punishment. "All Marines can't be punished the same, because we have to make sure that the NJP process is something that is still effective," McDade observes. NJP (or, non-judicial punishment) is a form of military justice authorized by Article 15 of the Uniform Code of Military Justice. NJP permits commanders to administratively discipline troops without a court-martial. Punishment can range from reprimand to reduction in rank, correctional custody (aboard ships only), loss of pay, extra duty, and/or restrictions.

The range of punishment is important to finding the just response to varying circumstances. Also, if the NJP is always the same, it can lose effectiveness. If a Marine knows precisely what the consequences will be for negative behavior, he can decide in advance if it is worth taking the risk. If the official response is tempered with a commander's evaluation or a subordinate leader's suggestion in determining a punishment, a Marine pondering misbehavior may think twice.

"In the eyes of the Marine Corps, looking at the Marine's record, the Commander may think that 'Okay. This Marine hasn't had any problems until now, so what's going on now?'" McDade explained. "They'll punish this guy, but it may not be as bad as a Marine who is not as good of a Marine, whose physical fitness is lacking, whose physical appearance is not all of that." Marine Corps leaders typically exercise discretion in doling out justice. They look at an individual's background and performance, impact of the decision on the unit, and even the effect on the entire Marine Corps in some serious cases.

Preserving and supporting justice can be a challenge, even for the most battle-hardened veteran Marine leader. McDade notes that it can be especially tough when justice must be applied to fellow Marines with whom he's served for a long time, some of whom he helped train. "You think that you know what you have at your disposal," McDade said. He continued:

> *"You think you know this guy, you think you know what he's capable of. You've had him since he was a young buck in training; you've seen him make tactical decisions. But when you get in combat, he's facing adversity, like rounds*

coming down range, or he sees one of his buddies get wounded or killed, everything you taught him, everything he saw, for that individual it kind of goes out of the window."

Obviously, training scenarios and real-world experiences often don't match up in picture-perfect alignment, and that discrepancy can sometimes be a morale killer. McDade remembers both the bad and good days in his multiple combat tours:

"There were times where we would go out and it would be a pretty good day. There might be a few firefights or whatnot, but no Marines would get hurt, and it would be a pretty successful day. And the Marines are motivated and they would be ready to go out again. Until the table turned, to where the day wasn't that successful, we still got the mission accomplished, but we lost a Marine. The next day you have Marines not wanting to go out. A lot of Marines were real quick to say that it's not fair, that it's not just, or whatever the case may be, for the enemy to be able to do the things that they were doing. They didn't operate under the Geneva Convention at all. They made up their own rules, they made up their own weapons, and they stalked us to learn our weaknesses."

McDade says that trying to motivate and encourage his Marines one of the most difficult aspects of combat for a leader. The key to success, he believes, is to explain the injustice in letting fellow Marines die in vain. "We sit here sitting on our hands, and we cry and we mourn them being gone, but the guy or the men that did it are still out there, wandering about, and trying to take away more of our

Marines," he said. It's painfully and emotionally wrenching for everyone involved and often difficult for a leader to break through in his attempts to maintain motivation. McDade explains:

> *"You know I hurt, I got the same lump in my throat, my heart is beating fast, I have chest pains, and I want to cry with them. And I want to get mine out. But I know when they see me do it, it's going to be one of those things where they say, 'Oh, it's Okay. He's crying. Now I know what we're doing is wrong, we're not going to do it.' So I had to swallow my feelings a lot and put them aside and tell the Marines that it's not just for us to sit up here, and let the enemy wander about and figure out a way to take more of us out, we got to find a way to stop them and get this thing over with so we can go home to our families and go tell the story of how brave our brothers were."*

However McDade managed to communicate with his Marines, it worked in some of the toughest times any of them will ever face. He kept his personal feelings in check because he understood his Marines had to do likewise if the unit was to succeed and survive. His men respected him because he took care of them, from the States throughout their tours of duty. He held his emotions in check, and that was enough to keep them motivated so they could load their magazines and get ready to get back and do the job.

"That was justice for me," McDade said:

> *"To know these guys respected me that much to follow me to the end of the world. I would say for us, while we were out there, we did a lot of missions. And I still don't think that I'm ready to say we found justice. But we tried."*[275]

Justice can be difficult to achieve—and difficult to mete out. It's a challenging aspect of leadership that affects the spirit of Marines, and of the Marine Corps itself. But effective leaders know that if they act fairly, don't play favorites, and avoid serving up justice with prejudice, they will be one step closer to keeping performance standards high and building morale. And, in doing so, they also help to foster loyalty, which is another key characteristic of NCO leadership.

Chapter 13

Loyalty

"There is a great deal of talk about loyalty from the bottom to the top. Loyalty from the top down is even more necessary and much less prevalent."
—George S. Patton, Jr. (1885-1945)

Among most people, loyalty is a fairly straightforward deal. We like something and so we stick with it, whether it's the brand of car we buy or the variety of breakfast cereal we chose. That's called brand loyalty and it usually evolves from personal and practical experience. We have good luck with Fords, so we buy Ford cars or trucks throughout our driving career if we can afford them. We grew up enjoying Lucky Charms, so the choice at the grocery store is easy. Rarely is any raw emotion involved in such simple choices. But that's not always the case when it comes to military service. Loyalty can get complicated.

In military organizations, loyalty is a valued trait, and it commonly is described as a two-way street. Loyalty is expected to flow up and down the organizational structure, known as the chain of command. A leader expresses loyalty to his subordinates by supporting their needs and ensuring their welfare in a number of ways. Subordinates express loyalty to that sort of caring leadership by positively and efficiently carrying out the leader's orders or instructions. That's the basics of loyalty from a Marine perspective, but an effective understanding of loyalty for leaders demands further examination.

Loyalty to a group can be defined most simply as staying in a group with a positive attitude, even when it is perceived that doing so benefits the group more than it benefits oneself.[276] Leaders have many tools at their disposal to increase loyalty, such as backing up their people when they are right, correcting them in private when they are wrong, and publically criticizing neither superiors nor subordinates. If that sounds like loyalty impinges or impacts on other leadership traits such as unselfishness, tact, and judgment, it's no wonder. Loyalty is the most common expression of aspects of all the Marine Corps leadership traits and characteristics. Those who get it express it through dedication and professional performance of duty.

Corporal Julie Nicholson found out about loyalty as an expression of behavior under stress during her deployment to Afghanistan in 2010, where she served as one of the key people keeping Bravo Surgical Company functioning in its crucial mission of saving Marines and sailors wounded in combat. Nicholson and a small group of Marines supported the Navy hospital corpsmen and doctors who constantly needed all sorts of supplies on short notice when lives were at stake.

Her job involved keeping supplies flowing to medical personnel at various Forward Operating Bases and to the primary surgical hospital at Camp Bastion in Afghanistan's Helmand Province. It could be routine, but Nicholson really came under pressure when emergencies happened, and those emergencies were abundant as combat grew increasingly violent in her area of operations. When the situation was urgent, Nicholson solved these short-fused situations by visiting supply warehouses, finding the required gear, and getting it out to the users in a hurry, often within hours. If narcotics or other medications were involved, they needed to be escorted to insure the shipment arrived safely and on time.

She did this job so well that she was selected for a number of honors, including certificates of commendation and a Navy and Marine Corps Achievement Medal. In the simplest terms, a Marine NCO like Nicholson demonstrated utter loyalty to her outfit and its mission in exemplary performance of duty. She also displayed loyalty to the wounded Marines and sailors who might have died without the proper life-saving medications or equipment. In addition, Nicholson's command showed loyalty to her through the decorations and commendations that were bestowed upon her.

Awards like these have meaning: they are a genuine expression not only of respect and appreciation, but also of loyalty—if they have been legitimately earned. That's where leaders must be careful with their expressions of downward loyalty. Unearned awards or recognitions can be counterproductive. Making everyone who labors in an organization Employee of the Month on some type of rotating schedule just to see that everyone has his moment in the limelight eventually renders the recognition meaningless. But when awards are truly earned and judiciously granted, they are expressions of loyalty flowing upward and downward in a unit. Leaders need to keep a sharp eye on that sort of thing.

"We work so hard every day to make sure that what needs to get done gets done," said Corporal Nicholson during an interview, "and for our seniors to notice and recognize that, it means a lot and it motivates you to want to keep going."[277]

While serving in Afghanistan, an important inspection was scheduled for Nicholson's supply section. It would mark the Marines involved as either on top of their game or in serious need of course correction. Her shop did well and received certificates of commendation. For Nicholson and her crew, it was motivating to know that their superiors actually knew that they were working hard and that they were willing to reward them for their efforts. Outstanding performance indicates loyalty up the chain of command; well-deserved recognition indicates loyalty down the chain. Nicholson understood how loyalty works and she got her first hints about it in recruit training.

U.S. Marine Cpl. Julie Nicholson checks out a laptop computer at "Shooters," a recreational center at the Transit Center at Manas, Kyrgyzstan, Feb. 19, 2010. Corporal Nicholson is a supply administrator and was headed to a forward operating base in Afghanistan. (U.S. Air Force photo/Senior Airman Nichelle Anderson/Released.)

After she completed high school, Julie Nicholson left her home in Ash Flat, Arkansas for Parris Island, where she excelled and was meritoriously promoted to private first class when she graduated from recruit training in April 2007. She went through Marine Combat Training in Camp Geiger, North Carolina, and Supply Administration School at Camp Johnson in North Carolina. She then was stationed at Camp Pendleton with the 1st Medical Battalion where hard work, perseverance and loyalty once again paid dividends. She was promoted to corporal and for the first time in her career had to start thinking about her personal stake in loyalty to her subordinates. She got a chance to see it all come together during an unusual experience for a Marine so new to Marine Corps leadership.

Nicholson was selected to go to Quantico, Virginia, for the 24th Executive Force Preservation Board, which discusses safety policies in the Marine Corps, such as motorcycle riding courses, suicide awareness, and proper protective equipment. In some ways, such policy-level meetings serve as examples of seniors being loyal to juniors by looking out for their welfare and listening to their insights. The juniors who provide those insights and opinions for the good of the organization are demonstrating their continued loyalty. Nicholson found the event to be encouraging:

"All these generals, the Assistant Commandant of the Marine Corps, Sergeant Major Kent [the Sergeant Major of the Marine Corps, the Corps' top enlisted man] were there just letting us know we have all done really

good on implementing the new courses and asking our opinions."

The event let the NCOs, that is, the very people tasked with implementing policy, know that their hard work and loyalty was making a difference and that it was appreciated at the highest levels.

Involving small-unit leaders in creating policies that affect the entire organization is effective and smart according to Nicholson. "They realized that they can make all these changes and do whatever they want," she said, "but it's really us NCOs that see what's happening and can make sure that the policies are actually implemented."

Developing loyalty often requires time, patience, and experience. However, although it may appear to be the case in some instances, the most loyal employee or Marine is not necessarily the one who has held the job longest. Some are simply marking time, with little or no interest in making valuable contributions to the organization. Those people may stick around simply because they have managed to cruise through by making few mistakes and avoiding making waves. Leaders need to determine if longevity comes from loyalty—or laziness.

To gain a clear understanding of loyalty and its application to leadership, it is useful for leaders to focus on their personal values and the values expressed in the ethos of the organization they serve. For example, some leaders might not believe that honesty has any particular value or that honest or negative criticism implies disloyalty to them personally. These are the "do as I say and not as I do" types who believe loyalty is expressed by an empty complaint box or team meetings during which no one dares utter a hint that things might not be going just fine. But the leader who expects that kind of behavior reflects a completely self-centered set of values and expects that loyalty need only be expressed in one direction: up.

Leaders who think like this are interested only in their own welfare and ignore the needs, concerns, and insights of the people who are expected to make things happen.

On the other hand, a leader who understands the value of loyalty down as well as up values the opinions of his subordinates, listens to their insights and suggestions, and acts on them, even if what's stated or implied indicates that the leader may not be doing the best job possible. It's a tough situation when a leader gets his toes stepped on by well-meaning subordinates or suffers a blow to his ego upon hearing that his performance could use improvement. But accepting that sort of criticism graciously is an expression of loyalty to the organization and to the subordinate unit. The key is communication.

Both formal and informal communication help generate ideas and improved performance. At Nicholson's current supply operation at Camp Pendleton, for example, the unit hosts a cookout or lunch at least once a month during which Marines can talk informally and away from the pressures of the job. Ideas are exchanged between superiors and subordinates, and egos are put on hold in order to keep them from getting in the way of effective communication. During informal situations like these, those who critique and those who listen are expressing loyalty to each other and to the mission of their outfit. It has proved a popular activity. It's no surprise to those who understand loyalty that solid ideas and valuable assessments spring from such occasions.

Does building loyalty depend on hosting bowling parties or barbecues? Not necessarily, though sometimes it can help, as it did with Nicholson's outfit. But the key is a consistent flow of information. Another helpful consideration that often helps build loyalty is healthy competition, which can provide a handy yardstick for leaders to assess loyalty.

The ability of a team to compete successfully against outside pressures is affected directly by the loyalty of its members. Those who stay actively engaged in the team bring with them crucial skills, knowledge, and values. When they put all those things into play, they are expressing a loyalty to their organization through the desire to be better than the competition. Also, their active commitment to the team in the face of temptation to, for example, relax, take the easy road, or simply let someone else contribute, actually increases the perceived value of the team both within the team and without. People outside the team start seeing it as a desirable team to which to belong, and this also strengthens the team in the eyes of the other team members.[278] This process then snowballs, as motivation to leave the team decreases and more members want to remain loyal to it. When a team member desires to stay and actively support the group in the face of potential outside benefits, he or she is expressing loyalty.

Leaders should all be so lucky in their organizations, but they often are not. At some point, all leaders are faced with unhappy subordinates, slackers, or people who generally could not care less about success. Leaders can take steps to get rid of such people if necessary, or they could express loyalty to those individuals by working with them to improve their attitude and performance.

Any experienced leader will tell you that coping with an unhappy subordinate is a challenge. In the Marine Corps, that challenge is unique in that the unhappy person generally doesn't have the option of just quitting. But an unhappy subordinate is difficult to live with and can affect the entire unit.

In an extreme case, an unhappy Marine with no sense of loyalty to his unit or to the Marine Corps in general can walk off the job. That's called desertion of one variety or another which is considered a crime under military law and carries stiff penalties. Instead, most disloyal Marines find other ways to express their displeasure or their animosity, and that presents a leadership challenge of its own. The easy way out is to kick the can down the road and transfer a problem Marine to another unit, thus making him someone else's responsibility. Short-sighted leaders, who are loyal only to their immediate command might exercise that option, but a leader who is loyal to the Marine Corps will not.

In Corporal Nicholson's unit, they do their best to keep their Marines motivated and loyal to their team. However, if a Marine doesn't like his assignment or the people he serves with, the standard operating procedure is to discuss the situation, find the root of the problem, and try to solve it. If that fails, there are options such as the Fleet Assistance Program which works a little like a lending library among Marine commands. Through this program, qualified Marines can be reassigned to temporary duty at a station that needs additional manpower. Nicholson is quick to emphasize that bad apples or chronic problems are not considered for the program. That would be kicking that problem down the road. But an otherwise good Marine who is just looking for something else more appealing to do during his enlistment may find the solution in a temporary assignment elsewhere. Loyal leaders often use the program to express their understanding and appreciation in the case of Marines who truly believe they are round pegs in square holes.

If a unit is temporarily short-staffed, it could ask to borrow a qualified Marine under the FAP process until a permanent replacement can be found. Leaders who take advantage of programs like this have a handy tool for expressing their loyalty to good men and women who just might do better with a change of venue. "One of my Staff NCOs would talk to that Marine, and let them know there was an opportunity to try something different for a few months, to take a break from the unit," said Nicholson. That Marine would go to another unit that needed him and perhaps find a better fit. If the leader is loyal to both the individual and the greater mission, this kind of transfer can work well.

Another aspect of effective communication impinges on loyalty down and often creates problems with loyalty up. It has to do with an understanding of and an appreciation for something sacred to all Marines: the mission. Much dissatisfaction in commands develops when there does not seem to be a solid connection between daily activities and the overall mission of the Marine Corps as a premiere fighting force. This dissatisfaction, which can breed disloyalty, stems from a failure to see a relevant value in what's being accomplished. Marine NCOs face this problem all the time and it's not uncommon among them to hear the old bromide: "When you're up to your eyeballs in alligators, it's hard to remember your mission was to drain the swamp."

If the work demanded of Marines every day seems menial, unimportant or unrelated to the larger mission of a fighting outfit training for—or actually engaged in—combat with mortal enemies, morale suffers. That can foment a sense of disloyalty to the unit. The loyal leader needs to jump into the situation with both feet at that point and start flexing some serious communication muscle.

It may take some creativity, but loyalty is inspired when a leader can effectively explain how what seems menial or trivial has a direct relation to the bigger picture. It may be tough to convince a young Marine that what results from a dreary day of tapping on a keyboard in an air-conditioned office has a direct effect on Marines closer to the sharp end of the bayonet, but doing so is crucial. Doing anything else, such as simply demanding that the puzzled Marine shut up and get back to the required drudgery, is counterproductive—and does nothing to build loyalty.

Good leaders need to show in a credible and reasonable way why the contributions of the team matter. Of course, that requires leaders to have a thorough understanding of what the bigger picture is.

Nicholson says the Marines in her unit know that what they do is important even though people doing administrative support jobs like supply often feel disassociated with more glamorous duties closer to the firing line. "We supply our surgical companies with all their gear," Nicholson said. That's an important task in support of combat life-savers but it often needs practical reinforcement. "Then they'll [the surgical companies] go out and do field operations and we go along so we can see they have what they need to function and do their jobs."[279] In cases like this, a simple field trip can be worth thousands of words.

Demonstrating a team's value can take some imagination. A Marine who was an NCO during the immediate post-Vietnam era admitted during an interview that he had to do some stretching on occasions when his fellow NCOs leading support units came to him for advice on how to improve motivation. He considered it questions from leaders who were trying to be loyal both to their Marines and to the Marine Corps as an organization, so he got creative.

On one occasion a young sergeant sought advice about his section of administrative specialists who were having some trouble seeing how what they did related to the Marines who were being celebrated for the job they did in the jungles and rice paddies of Southeast Asia. The sergeant knew the work his clerks did was important—even vital—but he was having trouble convincing the people who slaved over data entry machines in the rear areas.

Fortunately, the senior NCO had just completed a military history class during which unit diaries had been mentioned. Apparently, during some of the battles during World War II, reporting had been so poorly done that it took months, even years in some cases, to straighten out the records. Civilians had been told their family members were alive when they'd actually been killed in action on some Pacific island or, worse, Marines arrived home after their families had been told they were dead and buried.

"We made up a scenario about a family who were told their son died," he said:

"and all of the things the Marine Corps had to go through because of bad reporting. Good clerks could have kept it from happening in the first place, and could have helped make the corrections much easier as well. The presentation was a big hit."

A question remains: Could that young sergeant leading a team of Marine administrators have gotten his mission accomplished if he took a "shut up and stop whining" approach with the threat of official punishment in hand to anyone who hesitated or continued to complain? The answer is yes, but the lack of team loyalty to an important mission was bound to yield bare-minimum performance and shoddy work.

He would have become the sort of leader who issues this kind of order: "Just do it because the CO said we're supposed to do it. That's all you need to know." Questions about relevance would be answered like this: "It's important because I say it is. Just be quiet and get busy." If it's not obvious at this point, that's bad leadership and it demonstrates an approach to giving orders or passing instructions that can be deadly to loyalty up and down.

Unfortunately, the fear of being marked as the bad guy by delivering onerous requirements or other bad news is powerful. It is also a challenge to loyalty as a leadership trait. Let's take the case of an NCO who has just gotten a distasteful mission from her senior. She knows no one is going to like what the NCO is bound to say and will probably blame her for the upcoming tough times since she'll be the one insuring the orders to be carried out. She is afraid she is going to look bad in the eyes of her Marines. So, she arrives in the unit, gathers her subordinates, and says something like: "I don't like this any more than you do, but the CO says we've got to do it, so I guess we'd better get hot on it." That NCO has been mislead by ego into thinking she'll generate loyalty among her subordinates by presenting herself as a fellow sufferer, but she's wrong. She's also disloyal.

She's removed herself from the negative role in the short-term. It's not her issuing the orders. She's just passing the word from an evil slave-driver higher up in the chain of command and so is not to blame for the bad news. She's trying to convince her unit that the person who originally gave the order is the enemy, telling them that they should be loyal to each other and gut it out as a team. But that attitude won't last long, and most people will see right through the subterfuge. They know how things really work, and they expect that leader to be as loyal to her leaders as she expects them to be to her. Word gets around fast in a small outfit. It won't be long before the senior leader is questioning the loyalty of an NCO who is afraid to take responsibility and maintain motivation.

Corporal Julie Nicholson has become a devout believer in loyalty and she makes every effort to keep the traffic flowing on that two-way street. "When an order is given, I pass it on without any negative comments," she said. "Regardless of the order, I would never insult my superiors in front of my junior Marines. That's definitely how we show loyalty ... if we have problems we'd talk in private."

The loyal leader handles all negative feedback in private, whether it is directed at superiors or subordinates. "If a PFC in the shop was doing something wrong," Nicholson said. "I wouldn't call him out in front of everybody. I would wait until everything that is being done is over and I would pull him to the side and not embarrass him in front of the rest of the shop." Nicholson criticizes in private, but she is a big fan of praising in public. That's loyalty in the flesh, and it generates appreciation beyond commendation from her superiors.

Leaders with loyalty also offer benefits and opportunities for their subordinates when possible. Nicholson, for example, is using the tuition benefits offered by the Marine Corps to study criminal justice. She hopes someday to teach in a juvenile detention center. Once she knows when she will leave the Marine Corps and in what state she will live, she will pursue a teaching certificate. While that kind of benefit is beyond the power of the small unit leader, that leader's attitude and flexibility in allowing a subordinate to have the time to study and take advantage of the benefit is most certainly within their power.

"The tuition assistance I get by being a Marine is great," Nicholson says. "Everyone in my shop is pretty motivated to go to school. What's so important is having the support of our superiors for us spend the time. They really encourage us to go to school. We are all so grateful for that."[280] Her superiors are showing their loyalty to her, fostering her loyalty to them, and helping a Marine small-unit leader to improve her own skills and knowledge that she can bring back to the Corps.

Any organization with more than one team has more than one level of loyalty at play within the structure of the institution. With Marines, the levels can go from a fireteam to something as large as the United States itself. It is one of the reasons Marines take a formal oath on enlistment. Federal law[281] requires everyone who enlists or reenlists in the Armed Forces of the United States to take an oath that is filled with references to loyalty and for many it is their first hint at what is expected of them during their military service:

> *"I do solemnly swear (or affirm) that I will support and defend the Constitution of the United States against all enemies, foreign and domestic; that I will bear true faith and allegiance to the same; and that I will obey the orders of the President of the United States and the orders of the officers appointed over me, according to regulations and the Uniform Code of Military Justice. So help me God."*

This oath was purposefully written so that members of the military would be loyal to the Constitution first, rather than to any individual. Presidents come and go in their role of the Commander in Chief of the U.S. military, but the people who drafted that oath did so in the firm belief that the nation would never change its basic principles. It is to those principles that American service members pledge their lives and sacred honor. It is loyalty demanded by the promise of loyalty delivered to the citizen willing to make the sacrifice.

Chapter 14

Courage

"I would define true courage to be a perfect sensibility of the measure of danger, and a mental willingness to incur it."
—William Tecumseh Sherman (1820-91)

RESILIENCE?

Courage means standing up when life keeps knocking you down. It is holding on when every instinct is screaming for you to cut and run. It is one more step, one more try, on just one more day. Courage involves elements of both physical and psychological strength. In military circles, it is the undeniable hallmark of true leaders, and it is a function of all the other traits and attributes from initiative to knowledge to integrity.

A courageous leader faces the same physical risks and mental fears that his subordinates do, plus the pressure of making difficult decisions that may cost lives. That makes courage among military leaders not only admirable, but critical in crucial situations. Courageous leaders must demonstrate the kind of fortitude that will inspire courage in others, and that's a feat that often requires everything they've got in body, mind, and spirit. Marine leaders have given some serious thought to the subject. General Carl E. Mundy, the 30th Commandant of the Marine Corps, indicates courage is at the very heart of the Marine ethos:

> *"The heart of our Core Values, courage is the mental, moral, and physical strength ingrained in Marines to*

carry them through the challenges of combat and the mastery of fear; to do what is right; to adhere to a higher standard of personal conduct; to lead by example and to make tough decisions under stress and pressure. It is the inner strength that enables a Marine to take that extra step."[282]

The history of the Marine Corps is filled with examples of individual Marines demonstrating courage. From the earliest days of the American Revolution to the bloody battlefields of the Middle East and at every clime and place in between, Marines have been called on to show courage and have served as inspiration to generations of men and women held to the same high standards. The focus of many narratives used by Marines to define courage has often been on leaders like Corporal Walter Hiskett, who found himself shivering and shaking along with others of the 1st Marine Division at the infamous Chosin Reservoir in November 1950.

Hiskett joined the Marines from Chicago after a tough time surviving during the Great Depression. Hiskett thought he knew a little about hard times, sacrifice, and tenacity when he dropped out of school, scrabbled to support himself, and then joined the Marines for three years of steady pay and regular meals. Near the end of his enlistment, war broke out in Korea. He was on his way to a real test of courage that would make the mean streets of Chicago seem warm and nurturing.

Atop a snow and ice-covered hill overlooking a crucial pass that controlled traffic on the only road leading in or out of Chosin, Corporal Hiskett and his fellow Marines of the battered and under-strength Fox Company, 2nd Battalion, 7th Marines were ordered to hold the high ground and keep the Toktong Pass open at any and all costs while the rest of the division conducted a fighting withdrawal in the face of hordes of marauding Chinese. One word describes what Hiskett and the other Marines showed on what would become known in Marine legend and lore as Fox Hill: courage.

Hiskett was a fireteam leader in Fox Company's second platoon when he climbed the hill above the Toktong Pass. He knew how critical the company's mission was, and he knew the blowing snow and subzero temperatures would make the coming fight a nightmare. It was very much on his mind at about two o'clock on a frigid morning when he heard the sound of bugles. "What field music bugler is practicing at this hour of the morning?" was Hiskett's first thought, but it didn't take him long to understand the bugles were hardly a typical wake-up call. The next thing he heard was rifle and automatic weapons fire, but he decided that might be caused by green replacement troops up the road from Fox Hill who were suffering a case of shaky nerves. Rolling over in his sleeping bag, Hiskett had just about decided the fire was coming from the replacements at Hagaru-ri who were new to combat and shooting at shadows when he was violently disabused of that notion. The next thing he heard was, "Here they come!"[283]

The Chinese soldiers who hit Fox Hill that morning attacked in waves and outnumbered the Marine defenders by about ten to one. The first wave threw hand grenades. The second fired rifles and automatic weapons. The third wave was unarmed but they kept coming with orders to retrieve weapons from their fallen comrades and push on until they drove the Marines off Fox Hill.

It was chaos on the hill as Marines scrambled to respond to the swarming Chinese and keep their weapons firing in the freezing cold. In Fox Company's third platoon area, Private Hector A. Cafferata Jr., a big, burly Marine, threw off his sleeping bag and jumped into a nearby trench where he grabbed a pair of nearby weapons and started blazing away at the oncoming Chinese. "Cafferata stood up, completely exposing himself to the heavy fire of the enemy, and shot two M-1 rifles as fast as a wounded Marine could reload for him," Platoon Commander First Lieutenant Robert C. McCarthy reported. "Cafferata also grabbed one Chinese grenade, threw it out of the trench and pushed two others from the parapet."[284] Cafferata hit six of the Chinese threatening to overrun his position with the eight rounds in his rifle before being wounded in the right hand and arm by a grenade. According to his Medal of Honor citation, he continued to fight and "maneuvered up and down the line and delivered accurate and effective fire against the onrushing force, killing fifteen, wounding many more and forcing the others to withdraw." McCarthy credited Cafferata with "stopping what might have been a breakthrough just to the left of the Third Platoon."[285]

Corporal Hiskett, in a nearby section of the defensive line, was having similar problems. When the attack began, it was obvious that the Chinese were making an all-out effort to take Fox Hill. He sprinted to the nearest shallow foxhole, only recently chipped out of the frozen ground. It was filled with Marines already returning the wall of incoming fire, so Hiskett took what cover he could find behind some small dwarf pine trees and began to fire. "There was a great deal of noise," Hiskett remembered. "…the infernal bugles, Chinese and American shouts and curses…and the confusion of fighting in the dark. For the most part, I fired at silhouettes."[286] Hiskett stood erect so he could shoot over the men in front of him.[287]

One of the weapons used by the Chinese in the attack was the Thompson submachine gun, supplied to China under the lend-lease program during World War II. It was a man-killer of the Marines on Fox Hill. Hiskett quickly found out how effective that American weapon could be in close combat. "Suddenly I saw a Chinese soldier aim at me with a submachine gun." Hiskett remembered. "His first round hit me in the shoulder, knocking me to the ground while the rest of the burst snapped past my head. I was down for the count."[288]

Much more of that in the struggle for Fox Hill might have spelled disaster in the critical efforts to keep the Toktong Pass open, and that was on everyone's mind as the fighting intensified. The winding road that was the only access to re-supply the Marines and the only route for their withdrawal from the Chosin Reservoir could be easily cut by the Chinese if they controlled the pass. That would mean the 8,000 Marines moving to the rear would be trapped north of the pass.[289] And what that meant to the painfully wounded Hiskett was that he had to keep fighting for as long as he could. None of the Marines on that hill was suffering any delusions about the importance of their fight. They all knew the background that led them to the dangerous mission at Toktong.

When General Douglas MacArthur, Commander-in-Chief of the United Nations Command, achieved his brilliant end-around landing at Inchon in September 1950 which led to the recapture of the South Korean capitol at Seoul, he hoped to continue the push all the way to the Yalu River, the western border between China and Korea. MacArthur disregarded speculation that the Chinese would react to his offensive by coming into the war and shoved troops relentlessly northward in hopes of a quick, decisive victory. That put Hiskett and the rest of the 1st Marine Division in a long, strung-out line on the eastern side of the Chosin Reservoir in a rugged mountainous area moving slowly but steadily toward the Yalu as one of the worst Korean winters in history swept across the mountains and passes. And as predicted, it brought the Chinese Army flooding into the fight.

Units of the 1st Marine Division were stretched along sixty-five miles of the main supply route as night time temperatures dropped to thirty below zero. When the Chinese struck in force, it was clear to everyone in the Division that they would have to withdraw from their exposed positions and that meant the Main Supply Route was now their only escape from what could be a total disaster in North Korean territory. The back door leading out of the Chosin area had to be held open at all costs and the key to that was the Toktong Pass. That crucial mission fell to the 246 men in Fox Company of the 7th Marines, and they accomplished it over five days of bloody combat. At the end of that week, fewer than eighty of those men came off the hill alive and unwounded.

One of those men was Barry Jones who made it out of the Toktong fight and vividly remembered what *Time* magazine once described as "a battle unparalleled in U.S. military history."[290] Fifty years later, he still couldn't believe that he survived the fighting on Fox Hill when so many others did not. "It was so hard to see people get killed; people you grew up with," he recalled. "But still, they couldn't hit me ... thirteen months and they couldn't hit me."[291]

Barry Jones, a cold and exhausted Marine light machine gunner, climbs resolutely toward a ridgeline in Korea. (National Archives Photo (USAF) 342-FH-37885. Public domain.)

The Chinese enemy flooding into Korea and boring in on the trapped Americans at the Chosin Reservoir could and did hit a lot of other Marines. It's a point not lost on men like Jones who have had plenty of time to ponder the meaning of courage:

"Was I scared? Yeah, I was scared. Chicken. In the spring, a mortar round hit me dead on. I was holding a shelter half, folding it up, it hit between my legs, and my shelter half was just full of holes, and I had a few hunks of shrapnel in me, and that was it. Nothing to write home

about. The next day, when I went back up, a Corsair [airplane] did a power dive, and it sounded like that shell coming at me again. So I hit a trench. A trench that was being used as a latrine. I came out brown-side-out. So I got my hands dirty. But when I hit that trench, I wasn't afraid any longer. I'm always more afraid when I'm waiting and unsure of what to do than if I take action."

Jones had plenty of chances to take action on Fox Hill. The Marines holding that critical, frozen lump of high ground were attacked in strength by increasingly desperate Chinese soldiers every day and night for a week, but they held the ground due to the courage of men like Hiskett, Jones, and Sergeant John Henry, Fox Company's heavy machine gun section leader. He remembered:

"The guys were coming over the hill, and there were bodies all over the place, so the snow was blood-red, and it was freezing cold, it was so cold. I would say that the cold was a blessing to me, because I was wounded in so many different places I would have bled to death had it not been for the cold coagulating the blood."[292]

That bone-chilling cold sapped strength from the defenders on Fox Hill and added to the challenge in summoning the courage that each man required to resist rather than simply succumb to the brutal conditions. Jones, who retired after surviving Korea and a thirty-year career with the Los Angeles County Sheriff's Department, had frostbitten feet as did many of his fellow Marines at the Chosin Reservoir fight. But he's still walking and considers himself fortunate to have recovered from the frostbite as many others lost limbs because of the cold.[293]

The Chinese entry into the Korean War in 1950 wasn't the only surprise for the American fighting men that year. The bitter winter weather was something no one really understood or expected, especially the senior leaders responsible for seeing that the men on the lines had adequate clothing and equipment. They just didn't have the gear to combat the cold of a Korean winter. Boots were World War II-era "shoepacs" that were mostly intended to keep the feet dry but only caused problems for the men who tried them. The lace-up waterproof boots could not stand up to the bitter Korean winter, so "Mickey Mouse" boots—named for their large, comical appearance—were eventually issued but it was too late for Jones and a lot of other Marines.

"I got issued a pair of the new Mickey Mouse boots the day before I left Korea," Jones recalled. "Oh my God, they were beautiful. But it was after it was all over for me. I had to turn them back in the next day. At least I got to try them," he said. Nearly fifty years later, he has been diagnosed with peripheral artery disease. His legs are numb from just above the knees down to his feet. "I'm paying for it now. I got no feet. And my legs are gone."

When Fox Company reached the hill with their mission to hold it at all costs, company commander Captain William Barber, who would receive the Medal of Honor for his actions during the battle, considered not having his Marines dig in that night. It was late, and the men were tired. There was no sign of impending attack. But Captain Barber felt it best to consider all the possibilities, and he eventually gave the order to dig in, against the protests of his Marines.

Hiskett remembered the difficulty they had in carrying out that order. "The ground was frozen. We had entrenching tools and you could hear clanking entrenching tools pretty much into the night." Captain Barber did relent on one point: making the watch 30 percent instead of the more typical 50 percent, so that two men could sleep while one man remained awake on watch, rather than half of them needing to be awake and alert. Hiskett recalled his decision to take an early watch on a night that was nearly pitch black with visibility rapidly falling to zero:

> *"It was dark. It was dark, that first night. It was overcast and cloudy; it didn't really get bright, until next night. Then we were able to—we had the moonlight, then, but that night was very dark. But you could see where the Chinese were, because you could see the slugs coming from the muzzle of their weapon. So you could see a little bit of light flickering here and there. And you knew which direction to fire at."*

Hiskett and his fellow Marines were able to hold the line. When dawn appeared, the Chinese withdrew. Hiskett thought he was the luckiest man alive: "I thought I would be medevacked to a hospital, sleep on clean sheets, eat hot chow and take a bath, the first one in six weeks, but that never happened." There would be no evacuation for anyone on the hill. Fox Company was surrounded and cut off.[294] The Chinese encircled Fox Company on the hill in the same way they had surrounded the 5th and 7th Marines up the road at Yudam-ni, where they even managed to encircle the Division headquarters and one battalion from the 1st Marines to the east at Hagaru, where a vital airfield was operating.[295] Things were looking bleak for the Marines at Chosin and increasingly worse for the men holding Fox Hill.

On that hill, Barber had pitched two tents near the center of the perimeter, and that meager shelter was used as a collection point for a growing number of wounded men. Fox Company's Corpsmen did what they could for the wounded with nothing much more than field bandages and a dwindling supply of painkillers. Working by candlelight, "they changed the bandages, slipped men in and out of sleeping bags, warmed C-rations and melted the morphine syrettes in their mouths before injections," recalled First Lieutenant Robert C. McCarthy, one of the company's few surviving officers. The corpsmen had plasma, but they couldn't use it to replenish wounded men who had lost huge amounts of blood. It was frozen solid.[296]

Everyone faced death at nearly every moment on Fox Hill. Many of them became philosophic about it. "Death isn't the worst thing that can happen to you," Jones said. "I don't believe so. It's final. Staying alive there was worse. Death, it's all over, you don't feel it any more. At least I don't think so. That's what they tell me. When you're in a firefight, you don't run towards death ... you just do your job."[297]

Hiskett remembered he didn't have much time for thoughts like that or anything much besides just continuing to fight through his painful wounds:

"I could not fire a rifle with one good hand and arm, but I could throw a grenade. I'd hold it with my left hand and pull the pin and throw with my right. I don't remember

much of the fighting that night. When you get in one of those situations, you just react."[298]

For most of the men like Jones and Hiskett, it boiled down to a reliance on their training and a recollection of what they were told was expected of them as Marines. It's not an uncommon experience as other Marines testify in recalling similar situations on other battlefields around the world. Many of them recall having an NCO reassure them with words like "Remember your training and you'll do fine." It's one of the major goals of Marine training: instant, effective response in dangerous situations buoyed by regular reminders and examples of other Marines who have displayed huge measures of courage. The idea is to build into Marines an obligation to continue that legacy and live up to those traditions of courage in hard times.

The record indicates it works, but a number of Marines in Korea didn't have the benefit of that long, arduous training in the standard pipeline from boot camp to battlefield. In Korea, 8,500 Marines, including Barry Jones, never went to boot camp. They had to find a different well from which to dredge up the courage they displayed.

At the start of the Korean War, when the U.S. committed ground forces to the support of South Korea, General MacArthur almost immediately asked for additional units and for replacement personnel both to fill his understrength units and to replace battle casualties. The Marine Corps did their best to give MacArthur what he needed, but it wasn't easy. At the end of the World War II, active-duty manpower fell from its 1945 peak of nearly half a million men to about 74,000 troops by spring 1950. Also, there were no longer any Marine units of any significant size already in the Far East,[299] so the troops would have to be transported across the Pacific.

To get the needed number of Marines into the war, all of the ground reserve units as well as many of the reserve aviation units were called to active duty.[300] Because of this need for all the Reserve units, there was neither the time nor the resources to provide them all with recruit training.

Hector Cafferata, the Marine who helped stem the Chinese flood on Fox Hill, was one of those men rushed to Korea without benefit of recruit training. "All I know is that I'm proud that, even though I never went to boot camp, I can call myself a Marine."[301] No veteran would ever read his Medal of Honor citation and argue with that.

Jones was another who missed the boot camp experience:

"I had joined the reserves because I needed the money. We were poor. Four kids. So it was $3 a month, or something. That was big money to me. I was still in high school, too soon to join up in regular service. It was a good experience, for a little boy."[302]

But the experience didn't include the common denominator among Marines involved with a couple of rugged months at San Diego or Parris Island. Jones's reserve unit was called for service in Korea while he was still on the waiting list for boot camp. He got much of his training from combat veterans spotted among the replacements headed for the Far East aboard Navy ships. "I learned from the old guys," he said. "I learned on the ship. We fired off the fantail, both the M1 rifle and the machine gun. I think I shot a lot of albatross. There were still a lot left, though."

The men doing the instructing for the raw recruits often were World War II veterans and they had a significant impact on nervous young Marines like Barry Jones. "I remember one named MacMillan. I remember one guy tried to wake him up once and MacMillan near killed him." The training in transit was a lot less formal than that provided by Drill Instructors, but it was just as effective minus the formalities of drill and ceremonies. The focus was on combat and any other consideration was secondary. The NCO veterans recognized a need, stepped up, and brought their experience facing an earlier Pacific campaign to bear. They knew from hard experience that the young men heading for combat would need every advantage they could get in finding the courage to face what lay ahead of them in Korea.

Those NCOs who trained men like Jones and Cafferata also knew that the wellspring of courage lies in the spirit. They understood that physical courage is really a case of mind over matter in the most extreme conditions, when friends are dying right next to you every day. Hiskett testifies to the truth of that assertion.

He was struggling to keep going after being wounded in the initial attack on Fox Hill and fighting off waves of pain with morphine supplied by company corpsmen. His morale was being challenged with every hour that the Chinese kept coming and fellow Marines kept falling. In addition to his own wound, two of his best friends in the unit had been killed in the attack.[303] It was depressing, and he needed emotional courage more than anything else as the battle raged on and the Chinese closed in on Fox Hill. That emotional courage was a common commodity all across the firing line, especially on the second night of relentless combat with no possibility of replacements or reinforcements.

The 7th Marines' Commanding Officer, Colonel Homer Litzenberg, asked Barber if Fox Company could hold and keep the vital pass open. Barber's answer has become a watchword for courage in the Marine Corps: "We will hold, Sir." For the courageous, dwindling band of Marines on Fox Hill, there was nothing more to say.[304] Hiskett recalls many of them huddled in the tents among the wounded turned to other sources to find the courage required as the Chinese threatened to break through and overrun the company:

> *"When we were in the aid tent, we had gotten the word that the Chinese had broken through the 3rd platoon. And we could hear the voices of the Chinese issuing orders and so forth, getting closer and closer to the tent. And Lieutenant Schmidt, Larry Schmidt, who was our weapons platoon officer got the attention of all of us in the tent, and he said, 'Now look, you Catholic guys know your Rosary. Start saying it. And the rest of you guys just start praying and keep on praying until we get through this thing.' And he said, 'If they stick their head in the flap of the tent, just stare them in the eyes and show them you're Marines.'"[305]*

Hiskett was a tough kid from a tough background used to relying on himself, but that night he asked for special help in summoning his courage. He made a pledge: if he got off this hill alive he would serve God, forever, in any way he could:[306]

*"I didn't know what that meant at the time, but that was
the promise and the commitment that I made. And so we
kept on through the night, until finally we could see the
sunlight filtering through the bullet holes in the tent and
discovered that the 3ʳᵈ Platoon was able to stop them and
push them back. In the meantime, we had decimated the
two regiments that were attacking us at Fox Hill. As a
matter of fact, we used dead Chinese bodies to stack up in
front of our foxholes to provide some protection."*

There's no questioning the courage of the men who fought
for and held Fox Hill in Korea, but courage is an asset that
leaders need in other pursuits where what's at stake is less
final or drastic. It's a trait that can manifest in many ways,
including standing up to confront superiors. It's usually
called courage of conviction, and it is a tough thing to
demonstrate, especially in cases where a person is required
to resist and criticize common practice or firmly held
beliefs. Marine Lieutenant General Victor H. "Brute"
Krulak was often deemed a gadfly for his habit of standing
up to policy makers he thought were on the wrong track.
He was called a curmudgeon or a dinosaur by some in
government, but he spoke his mind throughout a long and
illustrious career. His philosophy regarding moral courage
is widely quoted in Marine Corps: "The essence of loyalty
is the courage to propose the unpopular, coupled with a
determination to obey, no matter how distasteful the
ultimate decision. And the essence of leadership is the
ability to inspire such behavior."[307]
   It is one thing to follow those guidelines when you are
wearing stars and another when you are struggling to show

courage farther down the chain of command. There is a human tendency to be cautious and courteous in criticism to keep from hurting the feelings of the people who need correction. Avoiding small hurts can often lead to more hurtful problems down the line for small unit leaders.

One example of this is the subordinate who is surprised when being fired, having had no indication that his performance was subpar, because his superior had never said anything for fear of hurting his feelings. The leader who failed to make initial correction or provide the necessary guidance—no matter how potentially hurtful at the time—has caused a larger problem by ignoring the initial situation. That leader may have reasons for it, but none of them involves courage. The appropriate behavior is to muster the courage it takes to be the bad guy and live with the resentment or anger that may entail. Another example is the leader who, when asked if future opportunities are available, tells the subordinate about wonderful prospects sure to come to fruition. This type of leader may be immediately popular, but when those opportunities never materialize, trust evaporates, and a long-term leadership failure has occurred. It takes courage to face the disappointment of your team.

Courage is never an easy commodity to find, whether it's disciplining a subordinate, standing up to superiors or facing swarms of charging enemy. Courage is situational; it lives in the moments when it is required by people who believe in themselves and in priorities beyond personal comfort and the risks of pain or failure.

The Marines of Fox Company understood that the courage they displayed on Fox Hill was vital to a larger mission, a crucial piece of a bigger puzzle that had to be solved even if doing so cost them their lives. They mustered the necessary courage from a variety of sources both traditional and instinctual, and they endured. It took massive displays of courage on the part of the leaders and equally huge, courageous responses from all the others to hold that ground and save the day at Toktong. That courage, mental and physical, is a big part of the reason the 1st Marine Division survived the battle of the Chosin Reservoir and brought all the Marines, living and dead, out of a deadly trap to survive and fight another day.

One of the first things Corporal Hiskett did when he finally reached safety, had his wounds treated in Japan, and spent some time pondering the fight on Fox Hill was to find a church where a military chaplain was holding services. Hiskett remembers what the chaplain said during the service:

> *"The content of his homily was pretty much, 'You know, you guys, when the chips are down and the going is rough and everything looks like it's pretty bleak,' the chaplain said, 'you have a tendency to make a lot of promises to God. You get back here and you're on clean sheets, eating hot food, your wounds are healing—you have a tendency to forget those promises.'"* [308]

Hiskett remembered the promises he made back at the Toktong Pass when he was seeking the courage to survive and keep his Marines in the fight. He went home, got out of the Marine Corps, and banged around doing construction work for a while. But something was eating at him, something that he kept recalling from his time on Fox Hill in Korea. It was the source of his courage and he believed it was time to repay some debts.

In 1962, he joined the Navy Chaplain Corps. And fifteen years after his service at the Chosin Reservoir, he volunteered to join his old outfit 2/7 heading for combat in Vietnam.[309] Hiskett understood courage, and he understood how to explain it to the sailors and Marines he now served as a chaplain. He stayed in Navy, reached the rank of Captain, and became the 9th Chaplain of the Marine Corps.

Barry Jones who knew Hiskett on Fox Hill, said it's a classic example of a man who has the courage of his convictions. "I remember him saying if he got out of there alive," Jones remembers from conversations with Hiskett in Korea, "that he'd live the rest of his life serving the Lord. And he did."

Conclusion

There is no measure of leadership except through action, whether on the battlefield or in a boardroom. The theory can be taught, but it is only through actions that the character of a person may be gauged. When someone is observed to be indifferent to assigned duties and responsibilities, then he is not ready to have more critical ones assigned to him, even if he has a reputation for good previous work or a charismatic personality. To lead, there must be passion for the work and for the people doing it. But many of the other personal qualities are variable.

There are as many types of leaders as there are types of people. They all have different origins, and they all have succeeded in different areas. Randy Burgess came from the Ozarks in southern Missouri, while Francini Fonseca was born in Brazil. Solomon Wren fought pirates on sailing ships; Carlos Hathcock fought the enemy in the jungles of Vietnam.

Some leaders may have been passed over early in their careers. Daniel Daly became one of the greatest heroes in the Corps, but he spent time early in his career in the brig. Aubrey McDade lost his father to a policeman's bullet and grew up in the inner city—the kind of kid society unfortunately tends to abandon. And yet these two men, and countless others, overcame their circumstances and some poor choices to go on to become leaders of Marines.

Leaders have very different personal quirks. Some won't touch alcohol, while others drink like fish. Some play sports; others love music. John Basilone liked to box. Lou Diamond loved animals. They can be either introverts or extroverts. Louis Cukela was blunt and outgoing. William Campbell March loved solitude.

Some were geniuses, but most were not. Donald MacGregor was still a teenager when he was a cadet captain in the ROTC, a journeyman machinist, and had completed two years of chiropractic college. John Basilone never finished high school. But they and others like them built their leadership abilities step by hard step. Julie Nicholson went from being an uncertain private to a confident NCO, presenting her ideas directly to Marine Corps generals. Walter Hiskett took his experiences on a frozen ridge in Korea and became Chaplain of the Marine Corps.

Although a very few of the best leaders may have been marked for greatness early in life, most weren't. They are human, and they all have their weaknesses. Not one perfect life is illustrated here. When Terry Anderson was a hostage in Lebanon, he was often frightened, lonely, and angry. Still, when the situation demanded it, he rose above these weaknesses and unselfishly led as he had been trained to lead. Although there are no perfect people, some can inspire others through near-perfect leadership.

Each of these men and women had a will to lead and a desire to bring out the best in others, in spite of their many differences. Tomorrow's leaders will be a mixed bag as well. Training for leadership does not mean removing diverse points of view and natural tendencies.

They had traits in common; all leaders do. By distilling their stories into the fourteen shared traits of Marine

NCOs, we can see those commonalities and learn to apply them in everyday circumstances far more mundane than combat, yet requiring leadership nonetheless.

Often, it is the inner qualities of a person rather than external marks of greatness that show us where true leadership can be found. Even though it is through their actions that we can observe their leadership, it is the internal workings of their minds and hearts that allow them to rise as inspirational leaders.

While there may be advantages to possessing great intellect, having great opportunities, and showing great intentions, these advantages are not enough if a leader is without character. And character is demonstrated through the leadership traits that Marine NCOs have demonstrated throughout history.

Of course, there are many qualities that are good to have in life that are not among the fourteen core leadership traits. Some leaders have no special gift for organization or technical skills, but they know how to find people with those gifts and bring out the best in them. Their gift is to see and understand the mission, judge it according to their resources and boldness, and unite their team so it works as one until the mission is accomplished.

The best leaders are genuine and don't try to be someone they are not. They are not actors, but they have a sense of the dramatic, know how to act to bring out the best when needed, and communicate in the most effective way both up and down the chain of command. They generally have a sense of humor, or at least, a sense of the absurd, and know when to use that as well.

They have courage. Followers will tolerate many things, but never will they follow cowardice or those who are timid at heart. Leaders have integrity, and they retain that through conflict and turmoil.

Leaders may vary in background, education, race, and in any number of other characteristics. But they all have the strength of the body when needed, the mind to learn and apply knowledge, and the spirit to lead others to victory.

Bibliography

Anonymous. "Adopted Valor: Immigrant Heroes Foreign Born Medal of Honor Recipients Sergeant Louis Cukela-WWI." *USCIS Monthly*, May, 2007.

———. "Army & Navy: Mortar Man," *Time*, February 22, 1943.

———. "Heroes: The Life & Death of Manila John." *Time*, March, 19, 1945.

———. "Men at War: Rescue," *Time*, June 4, 1951.

———. "Obituary of Lena Basilone." *Long Beach Press-Telegram*, June 16, 1999.

———. "Sgt. Maj. Carlton W. Kent." *Marines*, March 25, 2010.

———."Southwest Airlines - corporate personality." Business in the Community, November, 2005.

———. *The United States Army in Somalia, 1992-1994 CMH Pub 70-81-1*. Washington, D.C.: Center for Military History, 2002.

———"U.S. Will Spend $38.6 Million To Refurbish Port in Somalia." *UPI*, September 20, 1984.

"American Peace Commissioners to John Jay." Thomas Jefferson Papers, Series 1. General Correspondence. 1651-1827, *Library of Congress*. March 28, 1786.

Anderson, David A. "Effective communicative and listening skills revisited." *Marine Corps Gazette*, March, 2000.

Anderson, Gary. "Urban Warrior and USMC Urban Operations." *Marine Corps Warfighting Laboratory*.

Anderson, Terry. *Den of Lions*. New York: Ballantine Books, 1993.

Bartley, Robert. *The Seven Fat Years*. New York: Free Press, 1992.

Batalova, Jeanne. "Immigrants in the US Armed Forces." *Migration Policy Institute*, May 2008.

Bevilacqua, Allan C. "Next Time I Send Damn Fool I go Myself." *Leatherneck Magazine*, October, 2006.

Blum, Hester. "Pirated Tars, Piratical Texts Barbary Captivity and American Sea Narratives." *Early American Studies: An Interdisciplinary Journal*. 1.2 (2003): 133-158.

Boller, Paul F. Jr. *They Never Said It: A Book of Fake Quotes, Misquotes, and Misleading Attributions*. New York: Oxford University Press, 1989.

Boyd, Terry. "Young Marines learning to fight smarter and listen to local Iraqis." *Stars and Stripes*, September 24, 2006.

Brady, James. *Hero of the Pacific: The Life of Marine Legend John Basilone*. Hoboken, New Jersey: John Wiley & Sons, Inc., 2010.

Brofer, Jennifer. "Reserve Engineer Hopes to 'Spark Some Innovation' Against IED Threat." *1st Marine Logistics Group (FWD)*, September 28, 2010.

Brooks, Michael A. Jr. "Courage in the face of adversity." *Marine Corps Gazette*, September, 2001.

Broome, Norris C. "Policy is meant to be flexible." *Marine Corps Gazette*, August, 1972.

Buhl, Willard A. "Strategic Small Unit Leaders." *Marine Corps Gazette*, January, 2006.

Chambers, William and Robert Chambers, eds. *Chambers's Information for the People*, Volume 2. Edinburgh: W. and R. Chambers, 1842.

Chase, Eric L. "Courage on a Frozen Hilltop." *Marine Corps Gazette*, May, 2009.

Clark, George B. *Treading softly: U.S. Marines in China, 1819-1949*. Westport, CT: Praeger Publishers, 2001.

Clemons, Eric K. and Jason A. Santamaria. "Maneuver Warfare: Can Modern Military Strategy Lead You to Victory?" *Harvard Business Review*, April, 2002.

Cohen, William S. "Commencement Address." *United States Naval Academy*, May 26, 1999.

Cox, Chris W. "The Carolinas: It takes months to make a MEU special operations capable, but such capabilities are vital." *Leatherneck*, April, 2000.

Crilly, Benjamin. "1/5 mortarmen aim for proficiency and cohesion," *1st Marine Division*, December 10, 2010.

Cusack, John R. "A Psychiatrist Looks at Leadership Traits." *Marine Corps Gazette*, July 1986.

Department of Defense. *Active Duty Military Personnel Strengths by Regional Area and by Country*. September 30, 2010.

Diamond, Leland. "Letters." *Time*, January 10, 1944.

Dieckmann, Edward A. Sr. "Dan Daly: Reluctant Hero." *Marine Corps Gazette*, November, 1960.

Estes, Kenneth. *Handbook for Marine NCOs, Fifth Edition*. Annapolis: Naval Institute Press, 2008.

The Fighting Fourth of World War II. "Division History." Accessed November 3, 2010. http://www.fightingfourth.com/Maui.htm.

Freedman, David H. *Corps Business*. New York: Harper Collins, 2000.

Gaines, R.W. "Master Gunnery Sergeant Leland Diamond, USMC, Deceased." *Globe and Anchor!* March, 1956.

Garamone, Jim. "NCOs' Service Vital to Nation During Dangerous Time, Mullen Says." *American Forces Press Service*, April 22, 2008.

Gibbons, Floyd Phillips. *And they thought we wouldn't fight*. New York: Doran, 1918.

Gitman, Lawrence J. and Carl McDaniel. *The Future of Business: The Essentials*. Mason, OH: Cengage Learning, 2009.

Global Security. "Marine Corps History." Accessed August 22, 2010, http://www.globalsecurity.org/military/agency/usmc/history.htm.

――――. "Personnel End Strength." Last modified July 2010. http://www.globalsecurity.org/military/agency/end-strength.htm.

Goleman, Daniel. "War in the Gulf: P.O.W.'s; P.O.W.'s Now Told to Resist Cooperation to 'Best of Their Ability'." *New York Times*, January 24, 1991.

Gordon, Richard M. "Bataan, Corregidor, and the Death March: In Retrospect." *Battling Bastards of Bataan*. Last modified October 28, 2002, http://home.pacbell.net/fbaldie/In_Retrospect.html.

"Gunnery Sergeant Carlos N. Hathcock II: United States Marine Corps. 93 confirmed." Sniper Country, http://www.snipercountry.com/sniphistory.asp#Hathcock.

Harrison, Baron A. "Want to be an effective leader? Try being yourself." *Marine Corps Gazette*, August, 2000.

Hawkins, J. *Never Say Die.* Philadelphia: Dorrance & Company, Inc., 1961.

Heinl, Robert D., Jr. and John A. Crown. *The Marshalls: Increasing the Tempo.* Quantico, Virginia: HQUSMC Historical Branch, G-3 Division, 1954.

Henderson, Charles. *Marine Sniper.* Briarcliff Manor, NY: Stein and Day, 1986.

Hillman, Rolfe L. "Rediscovering Company K." *Marine Corps Gazette*, July, 1986.

"Hispanic Medal of Honor Nominees." Last modified January 22, 2008. http://www.hispanicmedalofhonor.com/nominees.html.

Historical Marker Society of America. "Gunnery Sargent [sic] John Basilone." Last modified September 3, 2009, http://www.historicmarkers.com/ca/80669-gunnery-sargent-john-basilone.

Holloway, Charles M. "Best damn mortar man in the Marines." *Marine Corps League*, Spring, 2005.

Hough, Frank O. "Dan Daly: He did spectacular things." *Marine Corps Gazette*, November, 1954.

Hunter, Stephen. "The Sniper with a Steadfast Aim." *Washington Post*, February 27, 1999.

Jolly, Vik. "First night we stacked 700 bodies." *Orange County Register*, September 16, 2010.

Jones, Jordan. "New solar-powered street lights are on in Kabul." *USFA-Central*, January 1, 2011.

Kaemmerer, T. J. "A hero's sacrifice." *1st Marine Logistics Group*, December 2, 2004.

Keene, R.R. "In the Highest Tradition." *Leatherneck*, July, 2007.

Klotz, Peter. "Politeness and Political Correctness: Ideological Implications." *Pragmatics*, 9:1.155-161, 1999.

Korb, Lawrence J. and Max A. Bergmann. "Marine Corps Equipment After Iraq." *Center for American Progress*, August, 2006.

Krulak, Charles. "Cultivating Intuitive Decisionmaking." *Marine Corps Gazette*, May, 1999.

———. "The Strategic Corporal: Leadership in the Three Block War." *Marines Magazine*, January, 1999.

Lansford, William Douglas. "John Basilone 's Last Battle." *Los Angeles Times*, May 3, 2010.

———. "The life and death of "Manila John"." *Leatherneck*, October, 2002.

Lantz, Gary. "White Feather." *America's 1st Freedom*. Archived from the original on September 27, 2007.

Laskin, David. *The Long Way Home: An American Journey from Ellis Island to the Great War*. New York: HarperCollins, 2010.

Leckie, Robert. *Strong Men Armed: The United States Marines vs. Japan*. New York: Random House, 1962.

Levine, John M. and Richard L. Moreland. "Group Reactions to Loyalty and Disloyalty." *Group Cohesion, Trust and Solidarity*, Volume 19, 2002.

Lind, William S. *Maneuver Warfare Handbook*. Boulder, Colorado: Westview Press, 1985.

London, Joshua. "Lecture On National Security and Defense." *Victory in Tripoli: Lessons for the War on Terrorism*, May 4, 2006.

Lopez, Steve. "An ex-Marine can run for us." *Los Angeles Times*, March 21, 2010.

Lowry, Richard. "Sgt. Rafael Peralta, American Hero." *National Review*, January 11, 2005.

Lydens, Peter F. "The Marines Who Never Went to Boot Camp." *Marine Corps Gazette*, January 2010.

MacLeod, Scott, et al. "Hostages: The Lost Life of Terry Anderson," *Time*, March 20, 1989.

Martin, Alexander. "The Magellan Star: Pirate Takedown, Force Recon Style." *U.S. Naval Institute*, September, 2010.

Martin, Iain C., ed. *The Greatest U.S. Marine Corps Stories Ever Told*. Guilford, CT: Lyons Press, 2007.

McCaffrey, James M. *Inside the Spanish-American war: a history based on first-person accounts*. Jefferson, North Carolina: McFarland and Company, 2009.

McCarthy, Dennis. "Memorial Day Patriots Give Thanks for the Courage of Men Like These: Honoring our Veterans." *Daily News*, May 30, 1999.

McCullough, Amy. "Ex-Navy star Adam Ballard booted from Corps." *Navy Times*, May 25, 2010.

Meid, Pat and James M. Yingling. *U.S. Marine Operations in Korea 1950-1953, vol. 5: Operations in West Korea*. Washington, D.C.: Headquarters USMC, 1972.

Moffett, Michael. "Oral History Interview: BGen Lawrence D. Nicholson." *United States Marine Corps History Division*, June 9 and August 17, 2010.

Moore, Crystal. "Sound Off." *Leatherneck*, September, 2007.

Moskin, J. Robert. *The U.S. Marine Corps Story*. Boston: Little, Brown and Company, 1992.

Mukden Prisoner Of War Remembrance Society (MPOWRS). http://www.mukdenpows.org/

National Museum of the Marine Corps. "Fate of the POWs Exhibit," *Korean War Gallery*, accessed November 14, 2010, http://www.virtualusmcmuseum.com/Korea_10.asp

Navy Department. *Annual Report of the Navy Department for the Year 1900*. Washington, D.C.: Government Printing Office, 1900.

Newcomb, Richard F. *Iwo Jima*. New York: Bantam Books, 1965.

Nilo, James R. "World War I: 75 Years Ago: The Battle of BLANC MONT." *Leatherneck*, October, 1993.

Nofi, Albert A. *Marine Corps Book of Lists*. Cambridge, MA: Da Capo Press, 1999.

Noonan, Peggy. "A Tragedy of Errors, and an Accounting: After a crash, the Marines set an example." *Wall Street Journal*, March 6, 2009.

Ormsby, Tom Jr. "Underwater Marines." *Marine Corps Gazette*, May, 1945.

Patton, Douglas E. "Enlisted PME Transformation." *Marine Corps Gazette*, February, 2006.

Perry, Tony. "Marines from Camp Pendleton who stormed pirate-held ship were combat veterans." *Los Angeles Times*, September 10, 2010.

Richardson, Herb and R.R. Keene. "The Corps' salty seadogs have all but come ashore." *Leatherneck*, November, 1998.

Robson, Seth. "A woman's touch: Engagement teams make inroads with Afghanistan's female community." *Stars and Stripes*, October 9, 2010.

Ruhl, Arthur. "Company K by William March." *The Saturday Review of Literature*, 1933.

Sanders, Gold V. "Push-Over Bridges Built Like Magic from Interlocking Parts." *Popular Science*, October, 1944.

Saucedo, Alfred. "Justice and the Art of Military Leadership." *Marine Corps Gazette*, August, 2010.

Schlosser, Nicholas J. "The Marine Corps' Small Wars Manual: An Old Solution to a New Challenge?" *Fortitudine* 35, 2010.

Schuon, Karl. *U. S. Marine Corps Biographical Dictionary: The Corps' Fighting Men What They Did Where They Served*. New York: Franklin Watts, Inc., 1963.

Schwartz, Barry. The Paradox of Choice: *Why More is Less*. New York: Harper Collins, 2004.

"2nd Battalion, 5th Marines Regiment," Official website, http://www.i-mef.usmc.mil/external/1stmardiv/5thmarregt/2-5/history/history_insignia.jsp

Scott, Stephen W. *Sergeant Major Dan Daly: The Most Outstanding Marine of All Time*. Baltimore: PublishAmerica, 2009.

Selby, John and Michael Roffe. *United States Marine Corps*. Oxford: Osprey Publishing, 1972.

Senich, Peter R. *The one-round war: USMC scout-snipers in Vietnam*. Madison, WI: Paladin Press, 1996.

7th Marine Regiment (REIN) "History of the 7th Marine Regiment." Last modified May 12, 2008, http://www.i-mef.usmc.mil/DIV/7MAR/history.asp.

Silver, Steven M. "Ethics and Combat." *Marine Corps Gazette*, November, 2006.

Simmonds, Roy S. *William March: An Annotated Checklist*. Tuscaloosa: University of Alabama Press, 1988.

Simon, Lee E. "Strategic Sourcing: Insights from Early Marine Corps Commodity Teams." *Defense AT&L*, May-June, 2006.

A Sixth Marine in the Great War. "Blanc Mont, Bloody Blanc Mont!" http://www.greatwarsixthmarine.com/blancmont.html.

Skelly, Anne. "Dan Daly: Legendary Marine 'Devil Dog'." *Leatherneck*, November, 1988.

Sloan, Bill. *Given Up for Dead: America's Heroic Stand at Wake Island.* New York: Bantam, 2003.

Smethurst, David. *Tripoli: The United States' First War on Terror.* New York: Presidio Press, 2006.

Smith, Graeme. "An oasis of relative calm in a sea of violence." *Globe and Mail*, April 7, 2009.

Stockridge, Frank Parker. *Yankee Ingenuity in the War.* Honolulu: University Press of the Pacific, 2002.

Stoffer, Alfred E. "Mortars: Weapons of Opportunity." *Marine Corps Gazette*, June, 1946.

Stoner, Bob. Ordnance Notes: M60 7.62mm Machine Guns (All Versions) http://www.warboats.org/stonerordnotes/M60%20GPMG%20R5.html

Streeter, Jennifer. "Using Tact in the Workplace." *Suite 101*, August 16, 2010.

Sullivan, Maura. "What a Marine Jet Crash Could Teach Wall Street." *Harvard Business Review*, March 11, 2009.

Tatum, Charles. "The Death of 'Manila John' Basilone." *Leatherneck*, November 1988.

———. "Prop Talk." Dinner Speaker at Commemorative Air Force. February 28, 2002. Transcript can be

accessed here:
http://www.goldengatewing.org/proptalk/speaker.cfm?ID=
37.

Thiran, Roshan. "Love and leadership go hand in hand:
Science of Building Leaders." *The Star*, February 19, 2011.

Thompson, Ben. "John Basilone." Accessed March 7,
2010. http://www.badassoftheweek.com/basilone.html

Tilghman, Andrew. "New CFT to simulate battlefield
demands." *Marine Corps Times*, April 21, 2008.

Tolbert, Frank X. "Diamond in the Rough." *Leatherneck*, August, 1943.

"Top Ten Snipers: Carlos Hathcock Places First." Military Channel,
http://military.discovery.com/technology/weapons/snipers/
snipers-01.html.

Tributes. "Obituary of Onnie Clem," August 9th, 2009.
http://www.tributes.com/show/86531004

USMC. "A Concept for Functional Fitness." *Deputy
Commandant for Combat Development and Integration*,
November 9, 2006.

United States Marine Corps. "Fighting to Belong." Last
modified December, 2010,
http://www.theusmarines.com/2010/12/.

———. *Marine Corps Mentoring Program Guidebook.*
Washington, D.C.: NAVMC DIR 1500.58, 2006.

———. *Mortars Gunnery.* Washington, D.C.: FM 23-
91, 2000.

———. *Operations & Readiness Command and Control.*
Washington, D.C.: MCDP 6 SSIC 03000, 1996.

———. *Small Wars Manual.* Washington, D.C.:
NAVMC 2890, 1940.

———. *Sniping*. Washington, D.C.: FMFM 1-3B, 1981.

———. "The 2010 United States Marine Corps Birthday Message." *Commandant of the Marine Corps General James F. Amos*, November, 2010, DVD.

———. *Warfighting*. New York: Doubleday, 1994.

U.S. Marine Corps History Division. "Who's Who in Marine Corps History: Gunnery Sergeant John Basilone, USMC, Deceased." Accessed February 17, 2010, http://www.tecom.usmc.mil/HD/Whos_Who/Basilone_J.htm.

———. "Who's Who in Marine Corps History: Master Gunnery Sergeant Leland Diamond, USMC, Deceased." Accessed February 13, 2003, http://www.tecom.usmc.mil/HD/Whos_Who/Diamond_L.htm.

Urwin, Gregory J.W. "How Marine POWs Hung Tough," *World War II*, May 8, 2008.

Van Buren, Mark E. and Todd Safferstone. "The Quick Wins Paradox." Harvard Business Review, January, 2009.

Vega, Lucas. "1/23 Marines train together to build combat skills." *USMC Press Release*, January, 27, 2011.

Zabecki, David T. "Paths to Glory: Medal of Honor Recipients Smedley Butler and Dan Daly." *Military History Magazine*, 2007.

Zarbock, Paul. "Transcript of Oral History of Walt Hiskett." *William Madison Randall Library*, University of North Carolina Wilmington, March 15, 2007.

Interviews

Anderson, Jerry (Sergeant, USMC), September 5, 2010.
Anderson, Terry (Sergeant, USMC), January 5, 2011.
Bayer, Robert (Sergeant, USMC), August 13, 2010.
Biggs, William (Sergeant, USMC), October 13, 2010.
Bunch, Sean (Sergeant, USMC), January 11, 2011.
Burgess, Randy (Sergeant, USMC), January 6, 2011.
de la Cruz, Jose (Sergeant, USMC), November 16, 2010.
Farnsworth, Freddie Joe (Staff Sergeant, USMC), December 5, 2010.
Fechner, Matthew (Sergeant, USMC), February 10, 2011.
Fonseca, Francini (Sergeant, USMC), January 31, 2011.
Ford, Joe (Lance Corporal, USMC), January 19, 2011.
Garcia, Robert (Corporal, USMC), November 2, 2010.
Giaretta, Brian (Staff Sergeant, USMC), October 13, 2010.
Grassl, Michael (Gunnery Sergeant, USMC), October 13, 2010.
Gutierrez, Faith (Corporal, USMC), October 13, 2010.
Hartsell, Bradley (Staff Sergeant, USMC), December 15, 2010.
Henderson, Charles (Chief Warrant Officer, USMC), February 12, 2011.
Ivy, Jonathan (Sergeant, USMC), October 13, 2010.
Jones, Barry (Corporal, USMC), August 24, 2010.

Kamaris, Keith (Staff Sergeant, USMC), October 13, 2010.

Kugler, Robert (Staff Sergeant, USMC), January 10, 2011.

MacGregor, Donald (Sergeant, USMC), December 2, 2010.

Mangio, Mark (Staff Sergeant, USMC), October 13, 2010.

Martinet, Michael (Gunnery Sergeant, USMC), October 13, 2010.

McDade, Aubrey (Staff Sergeant, USMC), February 14, 2011.

Nicholson, Julie (Corporal, USMC), February 11, 2011.

Ray, Glenn (Staff Sergeant, USMC), October 13, 2010.

Roxas, Mervin (Corporal, USMC), January 27, 2011.

Shuster, Mark (Major, USMC), July 6, 2010.

Silva, Jorge (Sergeant, USMC), December 2, 2010.

Stokey, Michael (Sergeant, USMC), August 14, 2010.

Weir, Richard (Sergeant, USMC), February 9, 2011.

Whitfield, Bruce (Gunnery Sergeant, USMC), January 23, 2011.

Wilhoit, Sheryl (Gunnery Sergeant, USMC), October 13, 2010.

Xiong, John (Sergeant, USMC), October 13, 2010.

Index

About the Author

Julia Dye, Ph.D. keeps the Entertainment Industry honest through technical advising and performer training, and helps Hollywood directors capture the realities of warfare in all aspects of the media. As a partner in the consulting firm Warriors, Inc., she was Weapons Master and provided training to Colin Farrell for the film Alexander and with the military advisory team, oversaw historical accuracy for the HBO series, The Pacific, among many other productions. Dye earned her doctorate in Hoplology (the anthropology of human conflict) from The Union Institute & University. Her business background includes venue producing at the 2002 Olympic Winter Games in Salt Lake City, Utah, where she handled all production needs for Figure Skating and Short Track Speed Skating.

[1] This version is from NAVMC DIR 1500.58, Marine Corps Mentoring Program Guidebook, February 13, 2006, 84. There have been other versions before and since, but this version outlines the idea of what an NCO is, what an NCO does, and what expectations the Corps has of its NCOs.

[2] While the Coast Guard is generally smaller than the Marine Corps, it serves under the Department of Homeland Security.

[3] Department of Defense, *Active Duty Military Personnel Strengths by Regional Area and by Country*, September 30, 2010.

[4] Steven M. Silver, "Ethics and Combat," *Marine Corps Gazette*, November 2006, 76.

[5] Jerry Anderson, (Sergeant, USMC) in discussion with the author, September 5, 2010.

[6] Robert Bayer, (Sergeant, USMC) in discussion with the author, August 13, 2010.

[7] Marines Staff, "Sgt. Maj. Carlton W. Kent," *Marines*, March 25, 2010.

[8] Jim Garamone, "NCOs' Service Vital to Nation During Dangerous Time, Mullen Says," *American Forces Press Service*, April 22, 2008.

[9] J. Robert Moskin, *The U.S. Marine Corps Story* (Boston: Little, Brown and Company, 1992), 805.

[10] Gold V. Sanders, "Push-Over Bridges Built Like Magic from Interlocking Parts." *Popular Science*, October, 1944, 94.

[11] *The United States Army in Somalia, 1992-1994, CMH Pub 70-81-1* (Washington, D.C.: Center for Military History, 2002)

[12] Randy Burgess, (Sergeant, USMC) in discussion with the author, January 6, 2011.

[13] Mark Shuster, (Major, USMC) in discussion with the author, July 6, 2010.

[14] Burgess, Interview.

[15] Shuster, Interview.

[16] Shuster, Interview.

[17] Pierce, *Warfighting*, v.

[18] Eric K. Clemons and Jason A. Santamaria, "Maneuver Warfare: Can Modern Military Strategy Lead You to Victory?" *Harvard Business Review*, April, 2002.

[19] Burgess, Interview.

[20] Joe Ford, (Lance Corporal, USMC) in discussion with the author, January 19, 2011.

[21] Lawrence J. Korb & Max A. Bergmann, "Marine Corps Equipment After Iraq." *Center for American Progress*, August, 2006, 7.

[22] Personnel End Strength, July 2010, http://www.globalsecurity.org/military/agency/end-strength.htm.

[23] Korb, "Marine Corps Equipment," 20.

[24] "U.S. Will Spend $38.6 Million To Refurbish Port in Somalia." *UPI*, September 20, 1984.

[25] Charles C. Krulak, "The Strategic Corporal: Leadership in the Three Block War." *Marines Magazine*, January 1999.

[26] Burgess, Interview.

[27] Shuster, Interview.

[28] Jennifer Brofer, "Reserve Engineer Hopes to 'Spark Some Innovation' Against IED Threat." *1st Marine Logistics Group (FWD)*, September 28, 2010.

[29] Brofer, "Spark Some Innovation."

[30] Donald MacGregor, (Sergeant, USMC) in discussion with the author, December 2, 2010.

[31] Tom Ormsby, Jr., "Underwater Marines," *Marine Corps Gazette*, May, 1945, 29.

[32] John R. Cusack, "A Psychiatrist Looks at Leadership Traits," *Marine Corps Gazette*, July 1986, 73.

[33] Robert D. Heinl, Jr. and John A. Crown, The Marshalls: Increasing the Tempo. (Headquarters: USMC, Historical Branch, G-3 Division, 1954), 80.

[34] "Division History," The Fighting Fourth of World War II, accessed November 3, 2010, http://www.fightingfourth.com/Maui.htm.

[35] MacGregor, Interview.

[36] "Division History."

[37] MacGregor, Interview.

[38] "Division History."

[39] Daniel Goleman, "War in the Gulf: P.O.W.'s; P.O.W.'s Now Told to Resist Cooperation to 'Best of Their Ability'," *New York Times*, January 24, 1991.

[40] Gregory J.W. Urwin, "How Marine POWs Hung Tough," *World War II*, May 8, 2008.

[41] Richard M. Gordon, "Bataan, Corregidor, and the Death March: In Retrospect," *Battling Bastards of Bataan*, last modified October 28, 2002, http://home.pacbell.net/fbaldie/In_Retrospect.html.

[42] J. Hawkins, *Never Say Die* (Philadelphia: Dorrance & Company, Inc., 1961)

[43] Urwin, "Hung Tough."

[44] Urwin, "Hung Tough."

[45] Bill Sloan, *Given Up for Dead: America's Heroic Stand at Wake Island* (New York: Bantam, 2003), 352.

[46] "Obituary of Onnie Clem," *Tributes*, August 9th, 2009. http://www.tributes.com/show/86531004

[47] Urwin, "Hung Tough."

[48] Mukden Prisoner Of War Remembrance Society (MPOWRS) http://www.mukdenpows.org/

[49] Urwin, "Hung Tough."

[50] Terry Anderson, (Sergeant, USMC) in discussion with the author, January 5, 2011.

[51] Scott MacLeod et al., "Hostages: The Lost Life of Terry Anderson," *Time*, March 20, 1989.

[52] Terry Anderson, *Den of Lions* (New York: Ballantine Books, 1993), 119.

[53] Anderson, Interview.

[54] Anderson, Interview.

[55] "Fate of the POWs Exhibit," Korean War Gallery, *National Museum of the Marine Corps*, accessed November 14, 2010, http://www.virtualusmcmuseum.com/Korea_10.asp

[56] Moskin, *Marine Corps Story*, 580.

[57] Anon, "Men at War: Rescue," *Time*, June 4, 1951.

[58] Moskin, *Marine Corps Story*, 582.

[59] Quoted in Pat Meid and James M. Yingling, U.S. Marine Operations in Korea 1950-1953, vol. 5: Operations in West Korea (Washington, D.C.: Headquarters USMC, 1972), 440.

[60] Urwin, "Hung Tough."

[61] James R Nilo, "World War I: 75 Years Ago: The Battle of BLANC MONT," *Leatherneck*, October 1993, 12.

[62] Roy S. Simmonds, *William March: An Annotated Checklist.* (Tuscaloosa: University of Alabama Press, 1988), xii.

[63] Simmonds, *William March*, xiii.

[64] "2nd Battalion, 5th Marines Regiment," Official website, http://www.i-mef.usmc.mil/external/1stmardiv/5thmarregt/2-5/history/history_insignia.jsp

[65] "Blanc Mont, Bloody Blanc Mont!" A Sixth Marine in the Great War, http://www.greatwarsixthmarine.com/blancmont.html.

[66] David H. Freedman, *Corps Business.* (New York: Harper Collins, 2000), 46.

[67] Bradley Hartsell, (Staff Sergeant, USMC) in discussion with the author, December 15, 2010.

[68] Hartsell, Interview.

[69] Frank Parker Stockridge, *Yankee Ingenuity in the War.* (Honolulu: University Press of the Pacific, 2002), 12.

[70] Bob Stoner, Ordnance Notes: M60 7.62mm Machine Guns (All Versions) http://www.warboats.org/stonerordnotes/M60%20GPMG%20R5.html

[71] Mark E. Van Buren and Todd Safferstone, "The Quick Wins Paradox," Harvard Business Review, January 2009.

[72] Hillman, "Rediscovering Company K," 45.

[73] Arthur Ruhl, "Company K by William March," The Saturday Review of Literature, 1933.

[74] Hillman, "Rediscovering Company K," 45.

[75] Anonymous, "Heroes: The Life & Death of Manila John," *Time*, March 19, 1945.

[76] Anon, "Heroes."

[77] William Douglas Lansford, "The life and death of "Manila John"," *Leatherneck*, October, 2002, 24.

[78] 7th Marine Regiment (REIN) "History of the 7th Marine Regiment." Last modified May 12, 2008, http://www.i-mef.usmc.mil/DIV/7MAR/history.asp.

[79] Moskin, *Marine Corps Story*, 272.

[80] Lansford, "Life and Death," 23.

[81] Lansford, "Life and Death," 23.

[82] Moskin, *Marine Corps Story*, 272-273.

[83] Robert Leckie, *Strong Men Armed: The United States Marines vs. Japan* (New York: Random House, 1962), 101.

[84] U.S. Marine Corps History Division, "Who's Who in Marine Corps History: Gunnery Sergeant John Basilone, USMC, Deceased." Accessed February 17, 2010, http://www.tecom.usmc.mil/HD/Whos_Who/Basilone_J.htm

[85] James Brady, *Hero of the Pacific: The Life of Marine Legend John Basilone* (Hoboken, New Jersey: John Wiley & Sons, Inc., 2010), 46.

[86] USMCHD, "Who's Who."

[87] The United States Marine Corps. *Warfighting*. (New York: Doubleday, 1994), 100.

[88] The United States Marine Corps. "A Concept for Functional Fitness." *Deputy Commandant for Combat Development and Integration*, November 9, 2006, 9.

[89] USMC, "Functional Fitness," 11.

[90] Andrew Tilghman, "New CFT to simulate battlefield demands." Marine Corps Times, April 21, 2008.

[91] Richard F. Newcomb, *Iwo Jima* (New York: Bantam Books, 1965), 94.

[92] Historical Marker Society of America. "Gunnery Sargent [sic] John Basilone." Last modified September 3, 2009, http://www.historicmarkers.com/ca/80669-gunnery-sargent-john-basilone.

[93] Ben Thompson, "John Basilone." Accessed March 7, 2010, http://www.badassoftheweek.com/basilone.html

[94] USMCHD, "Who's Who."

[95] Brady, *Hero of the Pacific*, 132.

[96] HMSA, "Basilone."

[97] William Douglas Lansford, "John Basilone's Last Battle." *Los Angeles Times*, May 3, 2010.

[98] Lansford, "Last Battle."

[99] "Obituary of Lena Basilone," *Long Beach Press-Telegram,* June 16, 1999.

[100] Charles Chuck Tatum, "The Death Of 'Manila John' Basilone," Leatherneck, November 1988, 58.

[101] Charles Tatum, "Prop Talk." (Dinner Speaker at Commemorative Air Force, February 28, 2002). Transcript can be accessed here: http://www.goldengatewing.org/proptalk/speaker.cfm?ID=37.

[102] Newcomb, Iwo Jima, 94.

[103] Leckie, Strong Men Armed, 440.

[104] Tatum, "Prop Talk."

[105] USMCHD, "Who's Who."

[106] David H. Freedman, Corps Business (New York: HarperCollins, 2000), xiii.

[107] Frank X. Tolbert, "Diamond in the Rough," Leatherneck, August, 1943, 16.

[108] Tolbert, "Diamond," 68.

[109] Tolbert, "Diamond," 16.

[110] Charles M. Holloway, "Best damn mortar man in the Marines," Marine Corps League, Spring 2005, 41-42.

[111] Tolbert, "Diamond," 16.

[112] Tolbert, "Diamond," 16.

[113] George B. Clark, Treading softly: U.S. Marines in China, 1819-1949 (Westport, CT: Praeger Publishers, 2001), 55.

[114] R.W. Gaines, "Master Gunnery Sergeant Leland Diamond, USMC, Deceased," Globe and Anchor! March, 1956.

[115] Tolbert, "Diamond," 17.

[116] Holloway, "Best damn mortar man," 40.

[117] Alfred E. Stoffer, "Mortars: Weapons of Opportunity," Marine Corps Gazette, June 1946, 47.

[118] United States Marine Corps, Mortars Gunnery FM 23-91 (Washington, D.C.: Headquarters Department of the Army, 2000), 2-1.

[119] Holloway, "Best damn mortar man," 40.

[120] U.S. Marine Corps History Division, "Who's Who in Marine Corps History: Master Gunnery Sergeant Leland Diamond, USMC, Deceased." Accessed February 13, 2003, http://www.tecom.usmc.mil/HD/Whos_Who/Diamond_L.htm

[121] USMCHD, "Who's Who in Marine Corps History."

[122] Lee E. Simon, "Strategic Sourcing: Insights from Early Marine Corps Commodity Teams," *Defense AT&L*, May-June 2006, 49.

[123] Lucas Vega, "1/23 Marines train together to build combat skills," *USMC Press Release*, January, 27, 2011.

[124] Norris C. Broome, "Policy is meant to be flexible," *Marine Corps Gazette*, August, 1972, 50.

[125] Anon, "Army & Navy: Mortar Man," *Time*, February 22, 1943

[126] Leland Diamond, "Letters," *Time*, January 10, 1944

[127] Michael Stokey, (Sergeant, USMC) in discussion with the author, August 14, 2010.

[128] Anon, "Army & Navy: Mortar Man"

[129] Benjamin Crilly, "1/5 mortarmen aim for proficiency and cohesion," *1st Marine Division*, December 10, 2010.

[130] David H. Freedman, *Corps Business* (New York: Harper-Collins, 2000).

[131] Douglas E. Patton, "Enlisted PME Transformation," *Marine Corps Gazette*, February, 2006, 15.

[132] Bruce Whitfield, (Gunnery Sergeant, USMC) in discussion with the author, January 23, 2011.

[133] Whitfield, interview.

[134] Holloway, "Best damn mortar man," 44.

[135] Charles C. Krulak, "The Strategic Corporal: Leadership in the Three Block War," *Marines Magazine*, January 1999

[136] Graeme Smith, "An oasis of relative calm in a sea of violence," *Globe and Mail*, April 7, 2009.

[137] Jordan Jones, "New solar-powered street lights are on in Kabul," *USFA-Central*, January 1, 2011

[138] Seth Robson, "A woman's touch: Engagement teams make inroads with Afghanistan's female community," *Stars and Stripes*, October 9, 2010.

[139] Francini Fonseca, (Sergeant, USMC) in discussion with the author, January 31, 2011.

[140] Terry Boyd, "Young Marines learning to fight smarter and listen to local Iraqis" *Stars and Stripes*, September 24, 2006.

[141] Nicholas J. Schlosser, "The Marine Corps' Small Wars Manual: An Old Solution to a New Challenge?" *Fortitudine* 35 (2010), 4.

[142] *Small Wars Manual*. NAVMC 2890. U.S. Marine Corps. 1940, declassified 1958. 30.

[143] *Small Wars*, 31.

[144] Michael Moffett, "Oral History Interview: BGen Lawrence D. Nicholson," *United States Marine Corps History Division,* June 9 and August 17, 2010, 6.

[145] Moffett, "Nicholson."

[146] Willard A. Buhl, "Strategic Small Unit Leaders," *Marine Corps Gazette*, Quantico, Jan 2006, 54.

[147] Freddie Joe Farnsworth, (Staff Sergeant, USMC) in discussion with the author, December 5, 2010.

[148] Farnsworth, Interview.

[149] Boyd, "Young Marines."

[150] Barry Schwartz, *The Paradox of Choice: Why More is Less* (New York: Harper Collins, 2004), 2.

[151] Lawrence J. Gitman and Carl McDaniel, *The Future of Business: The Essentials* (Mason, OH: Cengage Learning, 2009), 168.

[152] Baron A. Harrison, "Want to be an effective leader? Try being yourself," *Marine Corps Gazette*, August 2000, 38.

[153] William Chambers and Robert Chambers, eds, *Chambers's information for the people, Volume* 2 (Edinburgh: W. and R. Chambers, 1842), 222.

[154] T. J. Kaemmerer, "A hero's sacrifice" *1st Marine Logistics Group*, December 2, 2004.

[155] Richard Lowry, "Sgt. Rafael Peralta, American Hero," *National Review*, January 11, 2005.

[156] "Hispanic Medal of Honor Nominees," last modified January 22, 2008, http://www.hispanicmedalofhonor.com/nominees.html.

[157] Allan C. Bevilacqua, "Next Time I Send Damn Fool I go Myself," *Leatherneck Magazine*, October, 2006, 52.

[158] David Laskin, *The Long Way Home: An American Journey from Ellis Island to the Great War* (New York: HarperCollins, 2010), 202.

[159] Paul F. Boller, Jr., *They Never Said It: A Book of Fake Quotes, Misquotes, and Misleading Attributions* (New York: Oxford University Press, 1989), 41.

[160] Laskin, *The Long Way Home*, 202.

[161] Karl Schuon, *U. S. Marine Corps Biographical Dictionary: The Corps' Fighting Men What They Did Where They Served* (New York: Franklin Watts, Inc., 1963), 47.

[162] Bevilacqua, "Next Time I Send Damn Fool I go Myself," 52.

[163] Bevilacqua, "Next Time I Send Damn Fool I go Myself," 53.

[164] William S. Cohen, "Commencement Address," *United States Naval Academy*, May 26, 1999.

[165] Schuon, *U. S. Marine Corps Biographical Dictionary*, 47.

[166] Bevilacqua, "Next Time I Send Damn Fool I go Myself," 54.

[167] Jeanne Batalova, "Immigrants in the US Armed Forces," *Migration Policy Institute*, May 2008.

[168] USCIS, "USCIS Naturalizes Largest Number of Service Members Since 1955," last modified November 9, 2010, http://www.uscis.gov/portal/site/uscis/menuitem.5af9bb95919f35e66f6 14176543f6d1a/?vgnextoid= 628d8ef34e03c210VgnVCM100000082ca60aRCRD&vgnextchannel= a2dd6d26d17df110VgnVCM1000004718190aRCRD.

[169] USCIS, "Adopted Valor: Immigrant Heroes Foreign Born Medal of Honor Recipients Sergeant Louis Cukela-WWI," *USCIS Monthly*, May, 2007, 5.

[170] USMC, "Fighting to Belong," last modified December, 2010, http://www.theusmarines.com/2010/12/.

[171] Mervin Roxas, (Corporal, USMC) in discussion with the author, January 27, 2011.

[172] Steve Lopez, "An ex-Marine can run for us," *Los Angeles Times*, March 21, 2010.

[173] Lopez, "An ex-Marine."

[174] Robert Garcia, (Corporal, USMC) in discussion with the author, November 2, 2010.

[175] Garcia, Interview.

[176] USMC, "Fighting to Belong."

[177] Jorge Silva, (Sergeant, USMC) in discussion with the author, December 2, 2010.

[178] Jose de la Cruz, (Sergeant, USMC) in discussion with the author, November 16, 2010.

[179] Peter Klotz, "Politeness and Political Correctness: Ideological Implications," *Pragmatics*, 9:1.155-161 (1999): 157.

[180] Jennifer Streeter, "Using Tact in the Workplace," *Suite 101*, August 16, 2010.

[181] Edward A. Dieckmann, Sr., "Dan Daly: Reluctant Hero." *Marine Corps Gazette*, November 1960, 23.

[182] Dieckmann, "Dan Daly," 23.

[183] Dieckmann, "Dan Daly," 24.

[184] Dieckmann, "Dan Daly," 24.

[185] Stephen W. Scott, *Sergeant Major Dan Daly: The Most Outstanding Marine of All* Time (Baltimore: PublishAmerica, 2009), 116.

[186] Dieckmann, "Dan Daly," 24.

[187] James M. McCaffrey, *Inside the Spanish-American war: a history based on first-person accounts* (Jefferson, North Carolina: McFarland and Company, 2009), 3.

[188] McCaffrey, *Inside the Spanish-American War*, 4.

[189] David T. Zabecki, "Paths to Glory: Medal of Honor Recipients Smedley Butler and Dan Daly," *Military History Magazine,* 2007.

[190] Anne Skelly, "Dan Daly: Legendary Marine 'Devil Dog,'" *Leatherneck*, November, 1988, 62.

[191] "Annual Report of the Navy Department for the Year 1900," *Navy Department.* (Washington, D.C.: Government Printing Office, 1900), 1116.

[192] Skelly, "Dan Daly," 62.

[193] Dieckmann, "Dan Daly," 25.

[194] Skelly, "Dan Daly," 62.

[195] Scott, *Sergeant Major Dan Daly*, 31-32.

[196] Frank O. Hough, "Dan Daly: He did spectacular things," *Marine Corps Gazette*, November, 1954, 31.

[197] Skelly, "Dan Daly," 64.

[198] Skelly, "Dan Daly," 62.

[199] Klotz, "Politeness and Political Correctness," 158.

[200] Floyd Phillips Gibbons, *And they thought we wouldn't fight* (New York: Doran, 1918), 304.

[201] Albert A. Nofi, *Marine Corps Book of Lists* (Cambridge, MA: Da Capo Press, 1999), 181.

[202] Dieckmann, "Dan Daly," 27.

[203] Scott, *Sergeant Major Dan Daly*, 86

[204] Michael Grassl, (Gunnery Sergeant, USMC) in discussion with the author, October 13, 2010.

[205] Grassl, Interview.

[206] David A. Anderson, "Effective communicative and listening skills revisited," *Marine Corps Gazette*, March, 2000, 60.

[207] David Smethurst, *Tripoli: The United States' First War on Terror* (New York: Presidio Press, 2006), ii.

[208] "American Peace Commissioners to John Jay," March 28, 1786, *Thomas Jefferson Papers, Series 1. General Correspondence. 1651-1827*, Library of Congress.

[209] Joshua London, "Lecture On National Security and Defense," *Victory in Tripoli: Lessons for the War on Terrorism*, May 4, 2006.

[210] Moskin, *Marine Corps Story*, 36.

[211] Blum, Hester. "Pirated Tars, Piratical Texts Barbary Captivity and American Sea Narratives." *Early American Studies: An Interdisciplinary Journal*. 1.2 (2003): 133-158.

[212] Herb Richardson and R.R. Keene, "The Corps' salty seadogs have all but come ashore," *Leatherneck*, November, 1998, 19.

[213] Moskin, *Marine Corps Story*, 36.

[214] Smethurst, *Tripoli*, 160.

[215] London, *Victory in Tripoli*.

[216] Smethurst, *Tripoli*, 167.

[217] London, *Victory in Tripoli*.

[218] Moskin, *Marine Corps Story*, 37.

[219] London, *Victory in Tripoli*.

[220] Kenneth Estes, *Handbook for Marine NCOs, Fifth Edition* (Annapolis: Naval Institute Press, 2008), 94.

[221] John Selby, Michael Roffe, *United States Marine Corps* (Oxford: Osprey Publishing, 1972), 5.

[222] Estes, *Handbook for Marine NCOs*, 65-66.

[223] Charles Krulak, "Cultivating Intuitive Decisionmaking," *Marine Corps Gazette*, May, 1999, 18.

[224] Krulak, "Intuitive Decisionmaking," 19.

[225] Krulak, "Intuitive Decisionmaking," 19.

[226] Krulak, "Intuitive Decisionmaking," 19.

[227] Gary Anderson, "Urban Warrior and USMC Urban Operations," *Marine Corps Warfighting Laboratory*, 295.

[228] Krulak, "Intuitive Decisionmaking," 21.

[229] "Operations & Readiness Command and Control," MCDP 6 SSIC 03000 *U.S. Marine Corps Department of the Navy*, 1996, 63

[230] William S. Lind, *Maneuver warfare handbook* (Boulder, Colorado: Westview Press, 1985), 4.

[231] Lind, *Maneuver warfare handbook,* 5.

[232] MCDP 6, 65.

[233] Alexander Martin, "The Magellan Star: Pirate Takedown, Force Recon Style," *U.S. Naval Institute*, September, 2010.

[234] Chris W. Cox: "The Carolinas: It takes months to make a MEU special operations capable, but such capabilities are vital," *Leatherneck*, April, 2000, 28.

[235] Richard Weir, (Sergeant, USMC) in discussion with the author, February 9, 2011.

[236] Weir, Interview.

[237] Martin, "Magellan Star."

[238] Tony Perry, "Marines from Camp Pendleton who stormed pirate-held ship were combat veterans," September 10, 2010, Los Angeles Times.

[239] Matthew Fechner, (Sergeant, USMC) in discussion with the author, February 10, 2011.

[240] Fechner, Interview.

[241] Fechner, Interview.

[242] Weir, Interview.

[243] Charles Henderson, *Marine Sniper* (Briarcliff Manor, NY: Stein and Day, 1986), 30-31.

[244] Iain C. Martin, ed., *The Greatest U.S. Marine Corps Stories Ever Told*, (Guilford, CT: Lyons Press, 2007), 255.

[245] "Top Ten Snipers: Carlos Hathcock Places First," Military Channel, http://military.discovery.com/technology/weapons/snipers/snipers-01.html.

[246] United States Marine Corps, *FMFM 1-3B Sniping* (Washington, D.C.: Headquarters United States Marine Corps, 1981), 1-1.

[247] Charles "Bill" Henderson, (Chief Warrant Officer, USMC) in discussion with the author, February 12, 2011.

[248] "Gunnery Sergeant Carlos N. Hathcock II: United States Marine Corps. 93 confirmed," Sniper Country, http://www.snipercountry.com/sniphistory.asp#Hathcock.

[249] Stephen Hunter, "The Sniper With A Steadfast Aim," *Washington Post*, February 27, 1999.

[250] Gary Lantz, "White Feather," *America's 1st Freedom*, Archived from the original on September 27, 2007.

[251] Peter R. Senich, *The one-round war: USMC scout-snipers in Vietnam* (Madison, WI: Paladin Press, 1996), 372.

[252] Stephen Hunter, "The Sniper With A Steadfast Aim," *Washington Post*, February 27, 1999.

[253] Sniper Country.

[254] Henderson, Interview.

[255] Henderson, *Marine Sniper*, 213-215.

[256] Henderson, *Marine Sniper*, 217.

[257] Robert Bartley, *The Seven Fat Years* (New York: Free Press, 1992), 135, 144.

[258] Roshan Thiran, "Love and leadership go hand in hand: Science of Building Leaders," *The Star*, February 19, 2011.

[259] "Lay Off."

[260] "Southwest Airlines - corporate personality," *Business in the Community*, November, 2005.

[261] Peggy Noonan, "A Tragedy of Errors, and an Accounting: After a crash, the Marines set an example," *Wall Street Journal*, March 6, 2009.

[262] Noonan, "A Tragedy of Errors."

[263] Maura Sullivan, "What a Marine Jet Crash Could Teach Wall Street," *Harvard Business Review*, March 11, 2009.

[264] Sean Bunch, (Sergeant, USMC) in discussion with the author, January 11, 2011.

[265] Amy McCullough, "Ex-Navy star Adam Ballard booted from Corps," *Navy Times*, May 25, 2010.

[266] Henderson, Interview.

[267] Alfred Saucedo, "Justice and the Art of Military Leadership," *Marine Corps Gazette*, August 2010, 54.

[268] Saucedo, "Justice," 54.

[269] Aubrey McDade, (Staff Sergeant, USMC) in discussion with the author, February 14, 2011.

[270] McDade, Interview.

[271] Crystal Moore, "Sound Off," *Leatherneck*, September, 2007, 9.

[272] McDade, Interview.

[273] R.R. Keene, "In the Highest Tradition," *Leatherneck*, July 2007, 44.

[274] McDade, Interview.

[275] McDade, Interview.

[276] John M. Levine and Richard L. Moreland, "Group Reactions to Loyalty and Disloyalty," *Group Cohesion, Trust and Solidarity*, Volume 19, 2002, 203-228.

[277] Julie Nicholson, (Corporal, USMC) in discussion with the author, February 11, 2011.

[278] Levine and Moreland, "Group Reactions," 208.

[279] Nicholson, Interview.

[280] Nicholson, Interview.

[281] Federal statute in 10 U.S.C. § 502.

[282] Michael A Brooks Jr., "Courage in the face of adversity," *Marine Corps Gazette*, September, 2001, 58.

[283] Paul Zarbock, "Transcript of Oral History of Walt Hiskett," *William Madison Randall Library, University of North Carolina Wilmington*, March 15, 2007.

[284] Camp, "Frozen Valor," 17.

[285] Camp, "Frozen Valor," 17.

[286] Brooks, "Courage in the face of adversity," 58.

[287] Zarbock, "Transcript of Hiskett."

[288] Zarbock, "Transcript of Hiskett."

[289] Eric L. Chase, "Courage on a Frozen Hilltop," *Marine Corps Gazette*, May, 2009, 77.

[290] Dennis McCarthy, "Memorial Day Patriots Give Thanks for the Courage of Men Like These: Honoring our Veterans," *Daily News*, May 30, 1999.

[291] Barry Jones, (Corporal, USMC) in discussion with the author, August 24, 2010.

[292] HQMC DivPA, "The 2010 United States Marine Corps Birthday Message," from Commandant of the Marine Corps General James F. Amos, November, 2010, DVD.

[293] Vik Jolly, "First night we stacked 700 bodies," *Orange County Register*, September 16, 2010.

[294] Camp, "Frozen Valor," 17.

[295] Zarbock, "Transcript of Hiskett."

[296] Camp, "Frozen Valor," 18.

[297] Jones, Interview.

[298] Zarbock, "Transcript of Hiskett."

[299] Global Security, "Marine Corps History," accessed August 22, 2010,
http://www.globalsecurity.org/military/agency/usmc/history.htm.

[300] Peter F. Lydens, "The Marines Who Never Went to Boot Camp," *Marine Corps Gazette*, January 2010, 50.

[301] HQMC DivPA, "Birthday Message."

[302] Jones, Interview.

[303] Drury and Clavin, The Last Stand of Fox Company, 151.

[304] Chase, "Courage on a Frozen Hilltop," 78.

[305] Zarbock, "Transcript of Hiskett."

[306] Drury and Clavin, The Last Stand of Fox Company, 169.

[307] Brooks, "Courage in the face of adversity," 59.

[308] Zarbock, "Transcript of Hiskett."

[309] Camp, "Frozen Valor," 19.